THE

PUBLICATIONS

OF THE

Lincoln Record Society

FOUNDED IN THE YEAR

1910

PUBLICATIONS OF THE LINCOLN RECORD SOCIETY
OCCASIONAL SERIES

The Lincoln Record Society has been publishing editions of historical records since 1910 and it continues to do so by means of its Main Series and the Kathleen Major Series of Medieval Records. Through the Occasional Series the Society will now additionally issue monographs and similar publications relating to aspects of the history of the ancient county and diocese of Lincoln.

Also in this series

Steep, Strait and High: Ancient Houses of Central Lincoln
by Christopher Johnson

FARMING AND SOCIETY
IN NORTH LINCOLNSHIRE

THE DIXONS OF HOLTON-LE-MOOR,
1741–1906

RICHARD OLNEY

The Lincoln Record Society

The Boydell Press

© Lincoln Record Society 2018

All Rights Reserved. Except as permitted under current legislation
no part of this work may be photocopied, stored in a retrieval system,
published, performed in public, adapted, broadcast,
transmitted, recorded or reproduced in any form or by any means,
without the prior permission of the copyright owner

First published 2018

A Lincoln Record Society publication
published by The Boydell Press
an imprint of Boydell & Brewer Ltd
PO Box 9, Woodbridge, Suffolk IP12 3DF, UK
and of Boydell & Brewer Inc.
668 Mt Hope Avenue, Rochester, NY 14620-2731, USA
website: www.boydellandbrewer.com

ISBN 978 1 910653 05 0

A CIP catalogue record for this book is available
from the British Library

Details of other Lincoln Record Society volumes are available
from Boydell & Brewer Ltd

The publisher has no responsibility for the continued existence or
accuracy of URLs for external or third-party internet websites referred to in
this book, and does not guarantee that any content on such websites is, or
will remain, accurate or appropriate

This publication is printed on acid-free paper

Printed and bound in Great Britain by TJ International Ltd, Padstow

In memory of Philip Henry Gibbons
of Holton-le-Moor

CONTENTS

List of Plates	viii
Preface	xi
Map 1 Part of north Lincolnshire showing places mentioned in the text	xv
Map 2 Holton-le-Moor *c.* 1900, showing purchases by the Dixon family, 1741–1875	xvi
1 The Rural Context	1
2 The Grazier: William Dixon (1697–1781)	11
3 The Tenant Farmer: Thomas Dixon (1729–1798)	27
4 The Old-Style Farmer: William Dixon (1756–1824)	43
5 William Dixon as Philosopher and Philanthropist	61
6 The Man of Business: Thomas John Dixon (1785–1871), the Early Years	99
7 The Man of Property: Thomas John Dixon, the Later Years	123
8 The Ladies of Holton, 1871–1906	143
9 Farming and Landowning	155
10 Class and Community	171
Appendix 1: The Dixon Archive	185
Appendix 2: Genealogical Tables	191
1 Dixon	192
2 Parkinson	194
3 Roadley	195
4 Skipworth	196
Principal Sources	197
Index	203

PLATES

1 'Holton House': sketch by Amelia Margaretta Dixon (1833–1906, later Mrs Jameson Dixon), 1852.   79

2 The Hall, Holton-le-Moor, in April 2017.   80

3(a) 'Old Manor House from the Kitchen Garden': watercolour sketch by A. M. Dixon.   81

3(b) The park at Holton, looking north-east: watercolour sketch by A. M. Dixon.   82

4(a) The Church of St Luke, Holton-le-Moor: watercolour sketch by A. M. Dixon, 1852.   83

4(b) The Church of St Luke, Holton-le-Moor, in April 2017.   84

5 Farming accounts of Thomas Dixon (1729–1798), 1756–8.   85

6 Thomas Dixon's memoranda of the baptisms of his older children, 1756–64.   86

7(a) Holton Hall interior: Charlotte Roadley Dixon ill in bed – sketch by A. M. Dixon, 1854.   87

7(b) Holton Hall interior: ?the Morning Room – preliminary sketch by A. M. Dixon.   88

8 Miniature portrait, probably of Richard Roadley Dixon (1830–1871).   89

9 Record of daily farming operations, 1824.   90

10 Portrait of Thomas John Dixon (1785–1871) probably by A. Salomé, late 1850s.   91

11 Portrait of Mary Ann Dixon (1800–1885), wife of T. J. Dixon, by Benjamin Hudson, 1839–40.   92

12 Account of farm work kept by Robert Fanfield of Holton, 1822.   93

13 Account of farm work kept by William Wilson of Nettleton, 1822.   94

14(a–b) Religious notes and reflections by William Dixon (1756–1824), 1807.   95

15(a) Mrs Jameson Dixon seated in her bath chair.	96
15(b) Mrs Jameson Dixon's bath chair.	97
16 Double portrait of Charlotte Roadley Dixon and Amelia Margaretta Dixon by Benjamin Hudson, 1844.	98

PREFACE

'I at least have so much to do in unravelling certain human lots, and seeing how they were woven and interwoven, that all the light I can command must be concentrated on this particular web and not dispersed over that tempting range of relevancies called the universe.' Thus George Eliot in the fifteenth chapter of *Middlemarch*. She was of course being modest. Despite her narrow focus, she was well aware of the 'range of relevancies' that made her work of much wider – almost, one could say, of universal – significance.

Middlemarch is a great novel, not a work of local history. But local history too must always be conscious of the wider context. This book places at its centre the interwoven human lots that made up the history of one family, the Dixons of Holton-le-Moor, near Caistor in north Lincolnshire. By studying that family, however, it aims to contribute to a larger but hitherto somewhat neglected subject, the role of the middling sort or middle class in English rural society.

The Dixons were farmers, but they were no ordinary farmers, and in various ways their story is of unusual interest. Their progress as successful agriculturists can be traced over several generations, from the early eighteenth century to the late nineteenth. This was comparatively rare, but they were even less typical in that their farming profits were used over many years to build up a substantial landed estate in their native district. During this long period they thus had the opportunity to consolidate their social status in their own parish, in the area centred on their local market town, and in a wider district of north Lincolnshire. Their circle came to include not only prosperous fellow-farmers but other members of the rural or market-town middling sort – clergymen, merchants, doctors and solicitors. Finally, we can follow these developments with some precision, because the family was exceptional among farming dynasties in accumulating and preserving a very rich collection of legal, business and family papers, described further in Appendix 1.

The plan of the book reflects these family, local and cultural themes. Chapters 2 to 8 follow the fortunes of the Dixons generation by generation, but they are sandwiched between more widely ranging discussions.

Chapter 1 provides a general framework, and considers how factors of geography and locality influenced rural attitudes to questions of class and social hierarchy in the period under consideration. At the other end of the book Chapter 9 is concerned with farming and landowning, and examines how unusual the Dixons were in the way they first constructed and then retained their landed estate. Chapter 10 reviews the changing social position of the family against the background of a local community, or series of communities, that was also changing and evolving over the century and a half covered by this study. (That, again, was an interactive process of which George Eliot was well aware.)

* * *

What follows is a book that has been a long time in gestation. I became interested in north Lincolnshire farmers in the mid-1960s, when I was working on the nineteenth-century politics of the county for my doctoral thesis. Between 1969 and 1975 I was an archivist at the Lincolnshire Archives Office, where I first encountered the riches of the Dixon papers. As described in Appendix 1, I was involved in the deposit in the Archives Office of a large collection of papers from Holton, and was responsible for cataloguing it. In 1979 I contributed a volume on the nineteenth century to the *History of Lincolnshire* project that included a chapter on the 'middling sort'. But by then I was living in London, and finding it more difficult, despite the generous help of many kind friends in the county, to pursue my Lincolnshire researches. An earlier version of this book was completed in 1992, but its revision for publication in its present form has had to await the comparative leisure of my retirement.

A brief paragraph is necessary here to explain why the book stops at 1906, and why the family living at Holton today is called Gibbons rather than Dixon. In 1906 Amelia Margaretta Jameson Dixon died at Holton, the last of her family in the male line. She was succeeded in the Holton estate by a cousin on the female side, Thomas George Gibbons. His mother was by birth a Skipworth, and *her* mother had been the daughter of William Dixon of Holton (see Appendix 2, Tables 1 and 4). T.G. Gibbons took the name of Dixon on his succession, as did his son George Sperling Gibbons when he succeeded his father in 1937, but George's sisters retained the surname Gibbons. In 1970 George died unmarried, and the estate passed to a cousin, Philip Henry Gibbons, who however did not have to change his name to Dixon. On Philip's death in 2007 he was succeeded at Holton by his eldest son Jonathan, who lives and farms there today.

My greatest debt in the preparation and publication of this book is to successive members of the Dixon/Gibbons family. I was fortunate to get to know George Dixon in the late 1960s, and to be inducted by him into what at first seemed the mystifying ramifications of the Dixon and related pedigrees. After his death in 1970 I continued to benefit from the help and local knowledge of his sisters, the Misses Mary, Dora and Joan Gibbons. In their different ways they were all much attached to Holton and its history, and their memories, going back to their first arrival in the parish in 1907, were a valuable supplement to the written record. From the early 1970s I was fortunate yet again to get to know Philip Gibbons, who gave me much support and encouragement during the preparation of this book. I greatly regret that neither Philip nor Vanessa, who were so hospitable at Holton and later at Lincoln, did not live to see its publication, but I am most grateful to Jonathan Gibbons, now of Holton, for continuing the family's interest in the project, and for his permission to make use of the Dixon archive and other material in his possession.

My next most important collective debt is to my former colleagues at the Lincolnshire Archives Office (now Lincolnshire Archives), who supported my work on the Dixons, and to more recent members of staff who have patiently borne with my demands as a reader. I am also most grateful to my one-time colleague at Lincoln, Alan Readman, for lending me his detailed notes on major parts of the Dixon archive; to the late Rex Russell, for innumerable acts of kindness, including the redrawing of the maps in his inimitable style; to Brian Smith, who as Secretary of the Historical Manuscripts Commission granted me a period of study leave to complete a first draft of this book; to the Trustees of the Leverhulme Trust, for a grant towards my expenses in that connection; and to Brian Davey, who kindly let me see the text of his edition of Thomas Dixon's justice books prior to its publication by the Lincoln Record Society.

For access to material in public repositories I must thank archivists and librarians at The National Archives, the British Library, the British Library of Political and Economic Science, the National Archives of Scotland, Cambridge University Library (where the late Dorothy Owen made available documents from the Ely Diocesan Archives), Nottingham University Library, and the record offices of East Yorkshire, Essex, Herefordshire and Grimsby (formerly Humberside).

Over a period of many years a considerable number of individuals have responded to my enquiries, welcomed me to their homes or offices, or made available manuscripts in their possession. I hope that I have not forgotten any of them, and I am all too aware that it is now too late to thank several of them in person. But I can at least take this opportunity to record my gratitude to the following: Dr Rod Ambler, Mr Michael Anyan, Mr R.J. Betteridge, Professor Kenneth Cameron, Mr J. Charlton, Dr J. Menzies Clow (of The Tower House, Caistor), Mr and Mrs C. Cottingham (of Ewefield), Mrs Susan Ellis, Mr Paul Everson, Dr Mary Finch, Dr Jessica Gerrard, Mr Godfrey (of Searby), Mrs D. Hogg (of Oxgangs), Mr Jackson (of Holly House, Caistor), Mrs Mary Kerr, Emma, Lady Monson, Mr Christopher Ollard, Mr and Mrs Perry (of Stope Hill), Mrs Rex Russell (formerly Mrs Joan Mostyn Lewis), Dr Joan Thirsk, Miss Ruth Tinley, Mrs Pat Trottnow, Mrs Dinah Tyszka, Mr Roger Wallis, Commander Peter Wells-Cole, Mr Jonathan Wright and Mr J.M.B. Young.

I am grateful to the Trustees of the Tenth Baron Monson for permission to cite references to manuscripts in private possession.

My thanks are due to the Lincoln Record Society for publishing this volume, and to its Honorary General Editor Dr Nicholas Bennett for much help and advice. My gratitude also to Nicholas Bingham, Rebecca Cribb and their colleagues at The Boydell Press.

Dr Bennett is additionally responsible for all the photographic work for the illustrations. The photographs of a painting in family possession, of a recent view of The Hall, Holton-le-Moor and of images from the Dixon papers in Lincolnshire Archies are reproduced by kind permission of Mr Jonathan Gibbons, and the photograph of the bath chair, part of the Collections at Normanby Hall, by courtesy of North Lincolnshire Museums Service. Much appreciated help in connection with the photographs was given by Mr Gibbons, Evelyn Van Breemen at Normanby Hall and the staff of Lincolnshire Archives. Michelle Bird kindly edited the maps.

My wife Ruth has lived with this endeavour and has discussed many aspects of it with me over the years. It owes much to her patient encouragement. She also read the text before publication and made many valuable suggestions. Needless to say nobody but myself can be held responsible for any remaining errors of fact or judgement.

Richard Olney
June 2017

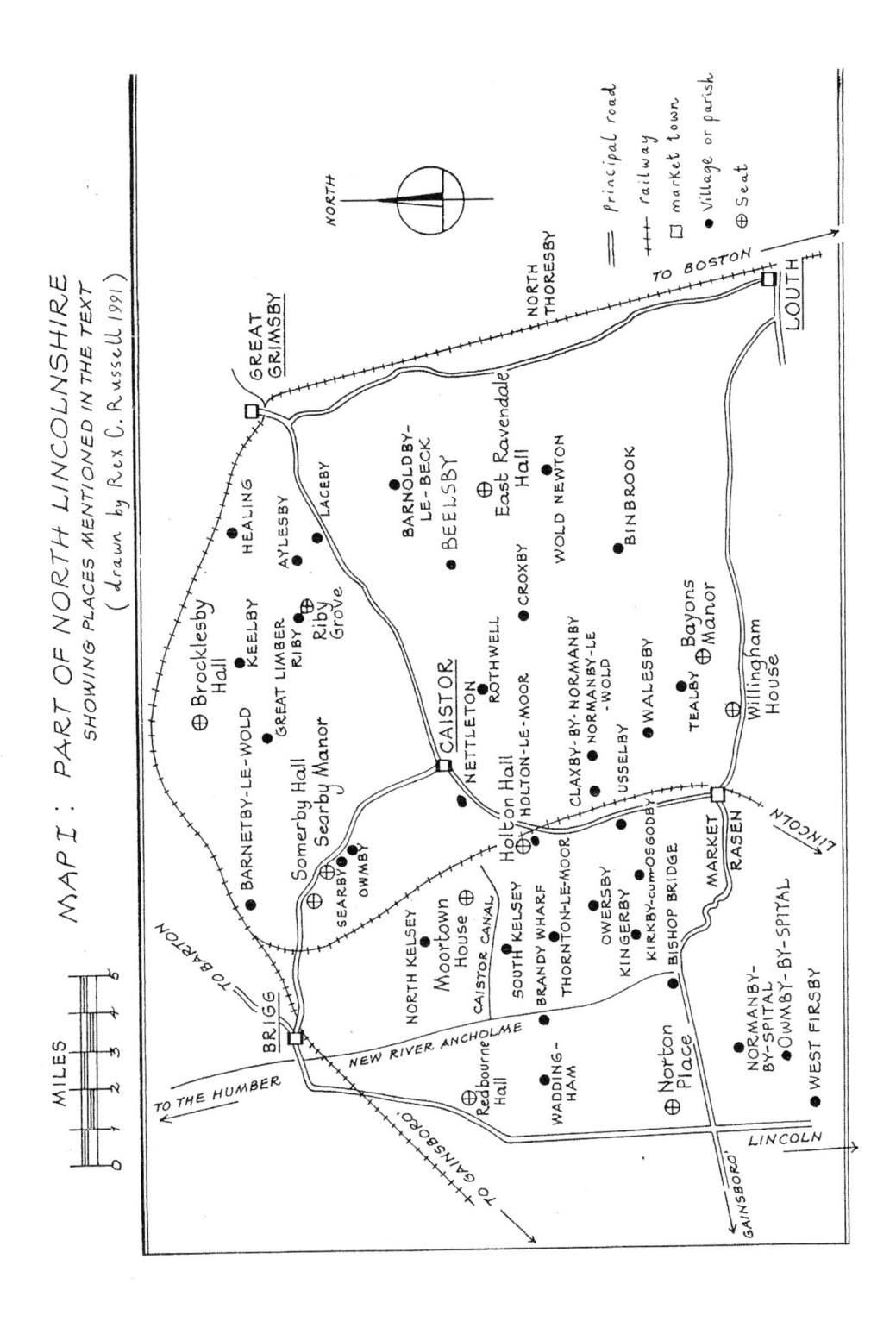

MAP I: PART OF NORTH LINCOLNSHIRE
SHOWING PLACES MENTIONED IN THE TEXT
(drawn by Rex C. Russell 1991)

MAP II

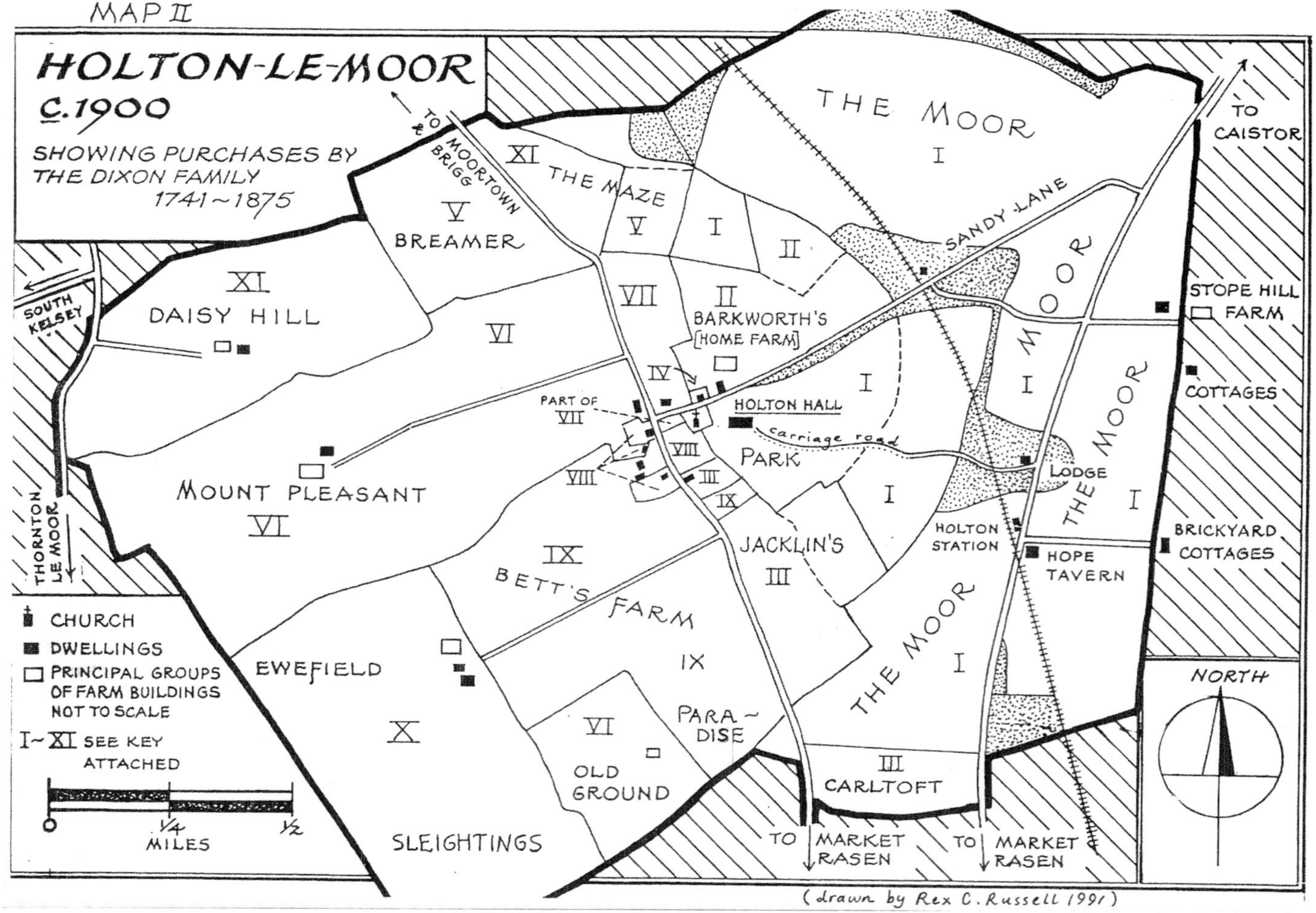

HOLTON-LE-MOOR
C.1900
SHOWING PURCHASES BY
THE DIXON FAMILY
1741~1875
TO MOORTOWN & BRIGG
TO CAISTOR
SOUTH KELSEY
THORNTON LE MOOR
THE MOOR I
THE MAZE
XI
V
I
II
BREAMER
V
DAISY HILL
XI
VI
VII
II
BARKWORTH'S
[HOME FARM]
SANDY LANE
MOOR I
STOPE HILL FARM
COTTAGES
IV
PART OF VII
HOLTON HALL
carriage road
I
PARK
VIII
III
IX
VIII
LODGE
THE MOOR I
MOUNT PLEASANT
VI
IX
BETT'S FARM
JACKLIN'S
III
I
HOLTON STATION
HOPE TAVERN
BRICKYARD COTTAGES
THE MOOR I
EWEFIELD
X
IX
PARA~DISE
VI
OLD GROUND
III
CARLTOFT
SLEIGHTINGS
TO MARKET RASEN
TO MARKET RASEN
NORTH
CHURCH
DWELLINGS
PRINCIPAL GROUPS OF FARM BUILDINGS NOT TO SCALE
I~XI SEE KEY ATTACHED
0
1/4
1/2
MILES
(drawn by Rex C. Russell 1991)

Key To Map II

I. Hall Farm and the Moor (including the site of the warren and later park and plantations) 1741. *c*.690 acres.
II. Barkworth's (later part of Home Farm), with right of free warren 1769. *c*.60 acres.
III. Jacklin's, with Carltoft 1775. *c*.77 acres.
IV. Noble's house and yard 1777. 5 acres.
V. Wiles's Farm (including the Breamer) 1789. 95 acres.
VI. Mount Pleasant Farm, with the Old Ground 1792. 319 acres.
VII. Broughton's house and closes 1814. 22 acres.
VIII. Bett's cottages and closes 1817. 9 acres.
IX. Bestoe's last property 1817. 170 acres.
X. Ewefield (later Yewfield) Farm 1840. 193 acres.
XI. Daisy Hill (or Grange) Farm 1875. 160 acres.

Notes

1. Properties I-IV were purchased by William Dixon (1697–1781), V and VI by William Dixon (1756–1824), VII-X by Thomas John Dixon (1785–1871) and XI by the Dixon trustees 1875.
2. The acreages for the earlier purchases are approximate. The figures given here total about 1,800 acres, compared with the total for the whole parish of 1,812 acres as given in the tithe award of 1838.
3. The Breamer (earlier Braymoor) was an area of rough pasture in the north of the parish. On the other side of the Moortown road was the Maze, an area of meadow later allocated to four separate holdings (I, II, V and IX). Carltoft (III), Paradise (IX) and Sleightings (X) were areas of pasture or meadow in the south of the parish. The boundary between Mount Pleasant (VI) and Ewefield (IX) may follow an earlier boundary between two arable open fields.

1
THE RURAL CONTEXT

In the eighteenth and early nineteenth centuries the Dixons were graziers and farmers. So were many of their relatives, and many of the neighbours with whom they regularly associated. They lived, moreover, in an age during which the acreage of productive land was extended, the investment of both capital and labour in the soil was greatly increased, the tools and implements of agriculture were transformed, the breeds of sheep and cattle were much improved, and new markets were opened up and exploited. It was, in short, an age of what has come to be characterised as agricultural revolution, or perhaps more accurately two revolutions, the first around the time of the Napoleonic Wars and the second during the mid-Victorian years of high farming.[1]

It was no coincidence that these two periods were ones of significant transformation in the fortunes of the Dixon family. This study will be concerned in part with when and how it made its money, and inevitably matters of agricultural practice will feature in its pages. But discussions involving horn and corn, cows and ploughs, will be kept firmly in their place, because this is primarily a piece not of economic but of social history. Its chief concern is with how the increasing wealth of the Dixons affected their standing in society. And because that society was to a considerable degree a *local* one, it will be necessary to consider questions of class and status in the context of the local community, or communities, in which the Dixons and their connections habitually moved.

In the eighteenth and nineteenth centuries a popular way of analysing English society was to divide it into three classes.[2] At the top sat the landed aristocracy, at the bottom were the labouring poor, and in between were a more heterogeneous group labelled the middling sort, or later the middle

---

[1] For a recent contribution to the ongoing debate on the Agricultural Revolution, see Susanna Wade Martins and Tom Williamson, *Roots of Change: Farming and the Landscape in East Anglia c.1700–1870* (British Agricultural History Society, 1999). For Lincolnshire, see T.W. Beastall, *The Agricultural Revolution in Lincolnshire* (Lincoln, 1978).

[2] For an excellent introduction to questions of class and hierarchy, see David Cannadine, *Class in Britain* (New Haven and London, 1998).

class or classes. As time went on this model became harder to apply to the expanding industrial towns, where the inhabitants appeared to be sorting themselves into only two classes, masters and men.[3] But it had a continuing appeal in the countryside, where it could be equated with the three most conspicuous social groups connected with agriculture – the large landlords, the tenant farmers and the farm workers. This mirrored the economic model according to which the landlords provided the fixed agricultural capital, the farmers the working capital, and the labourers the sweat of their brows. It also acknowledged the dominant position in rural society of the territorial magnates, who were to retain much power in their localities even after they had begun to lose their influence at the level of national government and politics.

This analysis, however, bore only an approximate correspondence to the realities of life, at least in rural north Lincolnshire. In some parishes the large landlord was an absentee, leaving the role of social leader to a clergyman or even a prominent farmer. In other parishes there was no large landowner at all, the soil being divided between a number of small proprietors. At the lower levels of rural society there were small craftsmen as well as labourers. But it is the variety within the farming class with which we are principally concerned here. There was a world of difference between the small farmer working his holding with the aid of family labour and the tenant of several hundred acres employing a labour force of fifteen or twenty men. Not all these substantial farmers, moreover, rented their holdings. Some might be owner-occupiers, others might own part of their holdings but rent the rest, while yet others might combine the roles of tenant farmer, owner-occupier and rentier landlord. The large farmer would naturally see his local society through different eyes from those of the small yeoman or peasant. The former might be closer in his attitudes to the landowning classes, although the interests of landlord and tenant were by no means identical.[4] And he might also grow apart in habits and sympathies from his men, excluding them from his house and his table.[5]

---

[3] This was less true of Birmingham or Sheffield, however, than it was of Manchester or Leeds (Donald Read, *The English Provinces c.1760–1960: A Study in Influence* (London, 1964), 36–7).

[4] Issues such as security of tenure or the right to kill game could cause friction between landlord and tenant, but on more essential issues, notably the maintenance of agricultural protection, they were more often in agreement (R.J. Olney, *Lincolnshire Politics 1832–1885* (Oxford, 1973), 198, 247–8).

[5] James Obelkevich, *Religion and Rural Society: South Lindsey 1825–1875* (Oxford, 1976), 51. It should be noted, however, that among farmers of over a hundred acres in

Besides which, there was more to the rural middle class than farmers and yeomen. The agricultural model excluded many entrepreneurs and professional people whose occupations were essential to the rural economy, even though they might not be directly engaged in agriculture. These occupational groups, like the farmers themselves, contained individuals of widely differing substance and status. The clergy ranged from poor curates to bishops, archdeacons and occupants of fat livings; the legal profession embraced down-at-heel clerks as well as partners in prosperous and influential firms of solicitors; doctors included both gentleman-physicians and tradesman-apothecaries; and men of business counted among them both petty village tradesmen and established merchants with wide-ranging connections.

Parsons, attorneys and merchants could of course belong to the urban as well as the rural middle classes. It is in fact urban historians who in recent decades have taken the lead in studying the transformation of an inchoate middling sort into a cohesive and confident local middle class.[6] The merchants and manufacturers of a place such as Leeds evolved their own distinctive culture. They laid great emphasis on the accumulation and transmission of wealth, and the chief agent of this process was the family. Marriage alliances were of great importance, and marriage partners were selected not just from other mercantile or industrial families but from the families of doctors, solicitors and clergymen. Also crucial were patterns of inheritance, with the middle class distinguishing itself from the landed gentry by its rejection of primogeniture and firm adherence to partible inheritance, treating all its offspring, male and female, fairly and equally.[7]

In contrast, much less work has been done on the rural middle class.[8] In studying the Dixons we shall consider how far they shared the characteristics

---

the Caistor union in 1851 households with living-in farm servants were still the rule rather than the exception. This was so even in some clearly well-to-do households.

[6] See, for instance, Peter Earle, *The Making of the English Middle Class: Business, Society and Family Life in London 1660–1730* (London, 1989); John Smail, *The Origins of English Middle-class Culture: Halifax, Yorkshire 1660–1780* (London, 1994); and, perhaps of most relevance to the present study, R.J. Morris, *Men, Women and Property in England 1780–1870: A Social and Economic History of Family Strategies amongst the Leeds Middle Classes* (Cambridge, 2005).

[7] Morris, *Men, Women and Property, passim.*

[8] A notable exception is the pioneering work of Leonore Davidoff and Catherine Hall, whose *Family Fortunes: Men and Women of the English Middle Class 1780–1850* (London, 1987) deals with the rural and market-town society of Essex and Suffolk as well as the urban society of Birmingham. For earlier periods, see David Rollinson, *The Local Origins of Modern Society: Gloucestershire 1500–1800* (London and New York, 1992), and J.R. Kent, 'The rural 'Middling Sort' in early modern England *c.*1640–1740', *Rural*

of the wider middle class, urban as well as rural; but it is clear that the rural 'bourgeoisie' was bound to differ from the urban in several key particulars. There was, to start with, the simple matter of space and distance. In the large towns merchants and manufacturers lived in close proximity to each other. They met for business and on public occasions in town centres, and increasingly lived in segregated suburbs such as Birmingham's Edgbaston, Leeds's Meanwood, or Leicester's Stoneygate. It was comparatively easy for them to achieve a kind of 'critical mass'.[9] In the countryside it was more difficult. Doctors, solicitors and merchants lived for the most part in county or market towns, but farmers spread themselves thinly over the rural parishes. Those who lived in villages were unlikely to be surrounded by their equals.[10] And those who lived in the middle of their farms, as many of the larger occupiers did by the nineteenth century, could be separated by a mile or more from their nearest congenial neighbours. Geographical factors were related to temporal ones: rural fortunes took longer in the making than urban ones, and the pace of life was slower, with country people tending to value continuity higher than change.

Attitudes to property could also be different. The merchant or industrialist might not be greatly worried by whether he owned or rented his premises, and if the latter his relations with his landlord were likely to be of a purely commercial nature.[11] In the country land had a social and political as well as an economic value, at least in some districts, and this had implications not only for landlord–tenant relations but also for those farming families that sought to invest their farming profits in the purchase of farms or small estates.

All this, however, begs the question of whether country people, including farmers, thought in terms of class at all, or whether they preferred to see their society in terms of status. Instead of aligning themselves with

---

History 10 (1999), 19–54. But some historians of the English middle class continue to write as though rural and market-town society had ceased to exist by the mid-nineteenth century, despite the fact that, as John Seed has pointed out, around 1850 over half the population of England and Wales was still living in places of under 2,500 inhabitants. (See John Seed, 'From "Middling Sort" to "Middle Class" in late eighteenth- and early nineteenth-century England', in M. L. Bush (ed.), *Social Orders and Social Classes in Europe since 1500: Studies in Social Stratification* (London and New York, 1992), 116.)

[9] 'Mere numbers are important. ... There are some thoughts which will not come to men who are not tightly packed.' (F.W. Maitland, quoted in Cannadine, *Class in Britain*, 21, n. 78.)

[10] This applied also to the parish clergy, who in many cases lacked family connection with their neighbourhoods of residence.

[11] Morris, *Men, Women and Property*, 15, 124.

other elements of the wider middle class, they preferred, it can be argued, to emphasise their local ties with their social superiors and inferiors. They saw themselves as occupying understood and accepted positions within a mutually supportive hierarchy. Hence the survival of deference in the countryside, a phenomenon that puzzled and exasperated the urban middle-class radicals of Victorian England, but that survived well into the twentieth century.[12] The papers of one member of the Dixon family give insights into the mind of an early nineteenth-century farmer who reflected on these issues, and Chapter 5 is based on this exceptionally informative source.

* * *

Deferential attitudes sit uncomfortably with class consciousness. They co-exist, however, more easily with ideas of community. As H.P.R. Finberg wrote in 1967, 'the local historian should concern himself not with areas as such, but with social entities.' These entities may be termed communities, and Finberg went on to define a community as 'a set of people occupying an area with defined territorial limits and so far united in thought and action as to feel a sense of belonging together, in contradistinction from the many outsiders who do not belong'.[13]

Armed with this definition, the local historian may be tempted to construct a 'social map' of a region or district, identifying the communities of which it was composed at a given date and defining the boundaries between them.[14] For various reasons, however, such a project would be most unlikely to succeed. For one thing, an individual might feel a sense of belonging to more than one group or entity – to a family in the first instance, and then to a village or parish, a part of a county, an entire county and even a whole nation. The boundaries of these communities could be depicted as a series of concentric circles. But that is a simple case. Some individuals might feel a sense of belonging to a number of *overlapping* communities, depending on whether they were thinking in terms of famil-

---

[12] Cannadine, *Class in Britain*, 122, 155, 158, 167.
[13] Quoted in Charles Phythian-Adams, *Re-thinking English Local History* (Leicester, 1987), 17. For a sociological version of the same definition, see Margaret Stacey, 'The myth of Community Studies', *British Journal of Sociology*, 20 (1969), 135.
[14] See, for instance, Keith Wrightson, 'Kinship in an English village: Terling, Essex 1500–1700', in Richard M. Smith (ed.), *Land, Kinship and Life-cycle* (Cambridge, 1984), 322. For an important mid-twentieth century anthropological study, see Marilyn Strathern, *Kinship at the Core: An Anthropology of Elmdon, a Village in North-west Essex in the nineteen-sixties* (Cambridge, 1981).

ial, social or economic connection. Another complication arises when one reintroduces the factor of social class. For the working-class villager one could argue that his community *was* the village, for the farmer that his primary community was the district centred on his nearest market town,[15] and for the country gentleman that, beyond his own house and estate, it was the county to which he felt the strongest allegiance. Again, however, the rural map was seldom a simple one. Not all the rural population was concentrated in nucleated villages; village did not always mean the same thing as parish;[16] a large farmer might have connections with more than one market; and in an extensive county such as Lincolnshire the lesser gentry might not play a prominent part in county affairs, or travel regularly to the county town. It may be useful, therefore, to conclude this introductory chapter by setting the Dixon family in its geographical context, and providing in outline the administrative framework within which, over the generations, it developed its communal networks and attachments.

* * *

Holton-le-Moor is a village and parish lying about seventeen miles north of Lincoln, and about fifteen miles south-west of Great Grimsby, in the County of Lincoln and the old administrative division of Lindsey (see Map 1). To the east rises the scarp of the Wolds, in the parishes of Claxby, Normanby, Nettleton and Caistor, and to the west, beyond the parishes of Thornton-le-Moor and South Kelsey, flows the river Ancholme. Holton itself has no chalk upland or alluvial lowland: much of it lies on a belt of sand, underlain by clay, that borders the Wold edge for several miles from near Market Rasen in the south to near Caistor in the north. Holton is a fairly small parish, of some 1,800 acres, and in 1842, when it was first described in print in a county directory, it had only about one hundred and fifty souls.[17] Not many years before that a substantial part of the parish had been moorland and rabbit warren, and the directory noted that it still had 'a moory appearance, abounding in ling, furze, etc', although by that date the common moor had been enclosed and some of the poorer soil

---

[15] According to Phythian-Adams, 'the impact of the local market centre cannot be underestimated [*sic*]' (*Re-thinking English Local History*, 23). See also G.J. Lewis, *Rural Communities* (London and Newton Abbot, 1979), for the spatial context of market towns as communal centres.

[16] K.D.M. Snell, *Parish and Belonging: Community, Identity and Welfare in England and Wales 1700–1950* (Cambridge, 2006), *passim*.

[17] William White, *History, Gazetteer and Directory of Lincolnshire* (Sheffield, 1842), 407.

was being covered by plantations. Roads connected the village to Market Rasen in the south, Caistor in the north-east and Brigg in the north-west, but none was a major highway. Brigg lay on the Ancholme, which was navigable northwards to the Humber, although, like the Trent further west, it restricted east–west communications by road. To get to the Trent at Gainsborough from Holton one went to Market Rasen, where one joined the turnpike road from Louth. Altogether Holton in the early nineteenth century was not exactly isolated, but it was some way off the beaten track.[18]

Holton was part of Walshcroft wapentake, which was centred on Market Rasen, but it was nearer to Caistor, which lay in Yarborough wapentake.[19] Although Holton had various connections with Market Rasen (and wapentakes still meant something in the early nineteenth century) it was Caistor to which it was principally linked, both economically and administratively. One of the most important links was ecclesiastical: although Holton was a separate parish it had no separate living. The 1842 directory (or rather its informant, presumably Thomas John Dixon) claimed that Holton was a vicarage, but in fact it was, and remains to this day, a chapelry of Caistor.

Caistor itself, perched on the Wold-side, was an ancient, indeed a Roman, settlement. A small town, of fewer than a thousand inhabitants in 1801, it yet retained some administrative importance, and served a wide district of the Wolds to the north and east as well as some of the lowland parishes to the west. In the early nineteenth century its population grew, and by 1842 it could be described as 'improving'. But it lacked a flourishing weekly market – Brigg and even Market Rasen were more important in that respect – and it was known locally mainly for its livestock, and above all its sheep, fairs. Nor was its age of improvement to last much longer. Brigg and Market Rasen had superior local communications, and this was accentuated when the railways arrived, linking Brigg and Market Rasen with the rapidly rising port of Grimsby but missing Caistor by more than three miles. On the other hand, the small size of Caistor, plus the fact that it was not dominated by one landed proprietor, meant that a few resident commercial and professional families could exert a significant influence over its affairs.

---

[18] Holton is less isolated today. The main road from Grimsby to Lincoln, the A46, passes through the parish. So does the railway line from Barnetby to Market Rasen and Lincoln, although Holton no longer has its own station.

[19] 'Wapentake' is the distinctively Scandinavian name used in Lincolnshire and other parts of the Danelaw for the divisions of the shire (known in other parts of England as 'hundreds').

Lincoln was less than twenty miles from Holton as the crow flew, but by road it was more like twenty-four, and in any case, despite its pre-eminence in the county as shire town and cathedral city, it did not exert the same pull as, say, Leicester did in Leicestershire. The environs of Holton were not studded with country houses whose inhabitants saw Lincoln as a social centre, and it was only when Holton and Lincoln were linked by a branch line of railway in the mid-nineteenth century that the business traffic between the two places noticeably increased. Of more importance in the latter connection had always been Holton's proximity to the prominent farming country of the Wolds.

In recent decades local historians have become interested in the concept of a natural region, or *pays*, that gives rise to a characteristic pattern of settlement and local culture.[20] A *pays* could correspond with a whole county. (A good fictional example would be the contrasted Stonyshire and Loamshire in George Eliot's *Adam Bede*.) Or it could represent a region within a county. In Gloucestershire, for instance, there are marked differences between the Cotswolds and the Severn Vale, differences which in the seventeenth century coincided with economic, social and religious divisions within the county.[21] In north Lincolnshire the Wolds, an extensive area of chalk upland, seem at first sight to be an excellent candidate for a *pays*.[22] Yet within the region there are differences between the southern Wolds, on the whole more picturesque and less dominated by large farms and estates, and the bleaker and more expansive landscapes of the area roughly north of the Louth-Market Rasen road.[23] It is this more northerly part of the Wolds, and particularly the area dominated by the Brocklesby estate and its leading farming tenants, that will feature more prominently in this narrative than other parts of the county within a similar distance of Holton-le-Moor. (See also Map 1.)

Commercially the region was partly defined by the market areas of six towns – Barton-on-Humber, Brigg, Great Grimsby, Caistor, Market Rasen and Louth – whose weekly markets were held for the most part on

---

[20] Alan Everitt, *Landscape and Community in England* (London and Ronceverte, 1985), 13–20, 41–59.
[21] Rollinson, *Local Origins of Modern Society*, 142 and *passim*.
[22] See, for instance, Charles Rawding, 'Society and place in nineteenth-century North Lincolnshire', *Rural History* 3, no 1 (April 1992), 59–85, and *The Lincolnshire Wolds in the Nineteenth Century* (Lincoln, 2001).
[23] Alfred Tennyson, brought up in the southern Wolds, regarded the country round Caistor as 'dreadful – barren rolling chalk-wolds without a tree', and Caistor itself as 'a wretched market town… at the limit of the civilised world' (Ann Thwaite, *Emily Tennyson: The Poet's Wife* (London, 1996), 105–6).

different days to accommodate the farmers and merchants who attended them. These towns shared the Wolds as part of their hinterland, but individually their market areas included lowland as well as upland, as we have already seen in the case of Caistor.[24] It was through Caistor, therefore, that Holton, though itself a lowland parish, was brought into contact with the wold parishes that extended behind the town in the direction of Grimsby and Louth. (The parishes nearer Barton lay too far to the north to make frequent contact practicable.)

* * *

The local historian, it has been said, can take a telescopic or a microscopic view of his or her subject.[25] In this case it would be possible to take a district of north Lincolnshire – always assuming that one had no trouble in defining that district – and examine the different communities within it. Alternatively one can begin with one place, even with one family, and work outwards, so to speak, from there. It is the latter course that has been followed here, and therefore it is to our chosen family and its local origins that we must now turn.

---

[24] Phythian-Adams in 1987 flagged up the need for 'a systematic study of the relationship of an important market town situated between, and mediating between, two contrasting *pays*, and the societies of the two *pays* in question' (*Re-thinking English Local History*, 11). In north Lincolnshire Louth, between the Wolds and the Lindsey Marsh, would be a good candidate for such treatment. In considering the Wolds as a *pays* it is also worth noting that both the Brocklesby estate and the Brocklesby Hunt, that important element in the social glue of the district, were by no means confined to the Wolds themselves.

[25] Phythian-Adams, *Re-thinking English Local History*, 18, 43–4.

2
THE GRAZIER: WILLIAM DIXON (1697–1781)

The landed connection between the Dixon family and Holton-le-Moor began in 1741, when William Dixon purchased the lordship of the manor, together with a farm in the parish.[1] At that date Holton would have seemed to a visitor a pretty unremarkable place. It was a fairly small parish, containing about eight farmsteads and about eleven cottages, with a total population of perhaps a hundred souls. Two of the farmsteads were outlying, but most of the buildings of the parish were concentrated in the village, around the church. None of these buildings was at all distinguished. The church was small and dilapidated, and there was no parsonage. The 'big house', built around 1600, had long since fallen into decay, and the house that went with the lordship, the old manor house, was no more than a farmhouse.[2] The farmhouses of the parish generally were all probably of the local one-storey type, with dormer windows in their thatched roofs,[3] and the cottages would have been of mud and stud.[4] The farmers were mostly resident, but the owners were mainly absentee, and this, together with the bad times that agriculture had recently experienced, would have given the place a somewhat shabby and neglected appearance. There was meadow land near the village, and west of the lane that ran from north to south through the parish there were sheep walks, but to the east of that road was a large area of sandy moorland, given over to bracken and furze and containing extensive rabbit warrens. Compared with today the parish had few trees, and the roads that led out of the village passed through

---

[1] Lincolnshire Archives, DIXON 1/A/1/12; and see also Map 2.
[2] See below, note 8.
[3] The only seventeenth-century stone-built house in Holton recognisable as such today is Broughton's. It was converted into three cottages in the early nineteenth century but retained its thatch until 1915 (information of Miss Joan Gibbons; letter to the author from Mrs M. Kerr of the Historic Buildings and Monuments Commission, 1984).
[4] The last cottage of this type survived as a store behind the Moot Hall until the late twentieth century.

similarly unromantic country. It was a poorly drained countryside, and probably difficult to negotiate during the winter months.

When William Dixon agreed the purchase in April 1741 he was described as a grazier, of Adlingfleet in Yorkshire. He was to pay a total of £685 5s for a farm of about 90 acres, including the old house near the church already mentioned, together with the lordship of the manor. The farm had a few years of a lease to run, and perhaps the chief value of the purchase was the 600 acres of manorial waste lying to the east of the Market Rasen–Brigg road. However, as will be seen, this did not give Dixon control over the warren that occupied part of that area, since the right of free warren had been alienated from the lordship of the manor some decades previously. The vendor was described in the agreement as William Bestoe, of Boston, gentleman. At first sight, therefore, this transaction might seem unlikely to have a great impact on the parish. Both the vendor and the purchaser were non-resident, and the purchaser would become only one of a number of owners in Holton. But in fact both parties had more local connections than is evident from the document; and in the longer term the transfer of this property from Bestoe to Dixon ownership marked the start of a process that was to transform the history of the parish. To understand its significance more fully it will be necessary to say something first about the Bestoe family and its property in Holton, and then about William Dixon's family and antecedents.

* * *

In 1578 Ottowell Bestoe, a well-to-do London merchant, acquired the lordship and parish of Holton-le-Moor from Sir Francis Ayscough, a prominent north Lincolnshire landowner who held the neighbouring parish of South Kelsey.[5] As with the purchase of 1741 the Bestoe acquisition could have been merely for investment purposes, but in this case too there was more to it than that. Bestoe was a wool merchant, and may have had regular dealings with the north Lincolnshire growers.[6] Furthermore, he himself may have originated from the locality. In a later herald's visitation his father is described as John Bestoe *of Holton*.[7] Ottowell died in 1584, but he may already have put in hand the building of a house at Holton for

---

[5] A confirmatory grant was obtained in 1611 (DIXON 1/A/1/1).

[6] For the provincial connections of London merchants at a somewhat later period, see Earle, *Making of the English Middle Class*, 44, 155. For the Trotmans, a seventeenth-century London family that retained close links with its Gloucestershire roots, see Rollinson, *Local Origins of Modern Society*, 107.

[7] College of Arms, C.23: visitation of Lincolnshire 1634.

his retirement. His son Nicholas certainly resided there, in a new house erected west of the main street around the turn of the seventeenth century.[8]

Nicholas Bestoe seems to have been not only a resident squire but also an improving one. It was probably he who enclosed the open arable fields and meadow land of Holton. The two largest fields lay to the west of the village, and Bestoe converted them to sheep pasture.[9] But he was not a depopulator, turning out families in order to let his sheep munch their way over what had been their small farms. When the common meadow known as the Maze was divided it was distributed among the larger farms of the parish, and about two-thirds of the moor remained rough pasture where the farmers and cottagers could graze their beasts and gather their fuel. The other third, however, was devoted to rabbits, whose value lay principally in their skins. They were kept in burrows, enclosed with sod walls topped with furze. Nicholas probably improved, or tried to improve, the productivity of this part of his estate, for later in the century the warren boasted 'a great lodge house and a little lodge house'.[10] He may also have rebuilt one or two of the farmhouses, and repaired and beautified the little church.[11]

This period of paternalistic prosperity was to last for less than half a century. Towards the end of Nicholas's life the family got into serious financial difficulties, principally it seems from having backed the wrong side in the Civil Wars, and from 1649 the estate was gradually broken up. The first two farms to be sold were Ewefield, in the south-west corner of the parish, and the grange farm later known as Daisy Hill in the northwest.[12] But this was not enough to salvage the family fortunes. On Nicholas's death the remaining properties descended to his grandson Richard (*c.*1638–1686), who disentailed them in 1659, presumably on his coming

---

[8] The foundations of the Bestoe house were discovered when Top Taylor's Yard was ploughed during the First World War. There is a drawing of its ground plan in DIXON 11/6/2. It had been abandoned as a Bestoe residence by 1673, when a deed describes the old manor house near the church as 'the capital messuage where Richard Bestoe now or lately dwelt', but it may have survived as a farm house into the eighteenth century (Nottingham University Library, Middleton Papers, Mi Da 67; Lincs. Archives, INV 200/171).

[9] See also Map 2. Ridge-and-furrow was still plentifully in evidence in Holton in the early twentieth century. The confirmatory grant of 1611 (see note 5 above) may have been obtained in connection with the enclosure. Owersby, just to the south, was partially enclosed in 1613.

[10] Middleton Papers, Mi Da 67.

[11] He gave a bell, and probably also a bible, in 1634 (DIXON 11/6/5).

[12] See also Map 2. The Daisy Hill conveyance of 1649 has not been traced, but a rental prepared for the new owner and dated 1 May 1649 survives among the Whichcote papers (Lincs. Archives, ASW 1B/77).

of age. He tried various ways of making ends meet, taking land into his own hands, bringing a small part of the Moor into cultivation, and moving from the house that his grandfather had built to the very modest old manor house near the church.[13] But in 1675–6 and again in 1683 he was obliged to part with various closes to his principal creditor Stephen Rothwell, the squire of the wold parish of Thorganby. In 1698 Richard's son William had to make another and final settlement with Rothwell, making over to him the farm later known as Barkworth's, three small holdings and the right of free warren. As already indicated, this gave Rothwell the right to manage and sell the rabbits, a business with which he was already familiar on his estate on the Wolds.[14]

William Bestoe was now left with the manorial rights of Holton, such as they were, that part of the Moor not given over to rabbits, the greater part of the village property, and some 300 acres of farm land. It was an estate small enough to be managed from a distance, and William went to live in Boston, where he married and set himself up in business. This state of affairs lasted for another forty years, but in 1738 William, now an old man, decided to divest himself of some of the Holton property. Benjamin Broughton, his leading tenant there, purchased his own house (still today known as Broughton's), with 18 acres near the centre of the village, plus some cottage property and small closes. But the £500 that this cost him exhausted Broughton's capital resources, and this gave William Dixon an opportunity to step in and become a landowner in Holton himself.[15]

* * *

William's earliest traceable ancestor in the male line is Robert Dixon, who died in 1593. He appears in a rental of 1576 as a tenant of the Monson family in Owersby, a large parish just to the south of Holton. He must have been a man of some substance and local standing, for he was churchwarden of Owersby in 1586, and when he died he left goods to the value

---

[13] See note 8 above.
[14] The 1698 conveyance gave Rothwell the 'liberty of getting and taking coneys or rabbits and fowle there with the tofts and little close there belonging, and licence… to make, plant and set burroughs for conies in and upon the moor called Holton moor or in any lands of [the] said William Bestoe by such engines… as are usual in such cases' (Mi Da 67). By this date the fact that a grant of free warren in Holton had been made in 1331 had been lost sight of, but it was also possible to establish the right prescriptively by long usage.
[15] DIXON 1/A/4/1–4.

of £174.[16] Another Robert, in all likelihood his son, was born at Owersby in 1570 and served as churchwarden there in 1597, but early in the next century he moved to the neighbouring parish of Usselby, where he died in 1627. His own son William was born at Owersby in 1603, but settled at Middle Rasen, again not far away, where he married and worked as a cooper. Dying in 1662, he left his cottage to his wife and then to his sons John and Thomas. He was a smallholder as well as a craftsman, since his probate inventory notes two horses, ten sheep, a cow, a pig and chickens, but his goods were valued at a total of only £12 3s.[17]

Thomas Dixon seems to have done better than his father. He farmed from around 1685 to 1690 at West Rasen, where he married the miller's widow, from around 1690 to 1694 at nearby Newton-by-Toft, where he married another widow, and then at Kirkby-cum-Osgodby (south of Owersby and north of the Rasens). It was at Kirkby that Thomas married his third wife, Ann Waltham from Wragby, and at Kirkby that William, the purchaser of Holton, was born in 1698, the only son by any of his father's marriages to survive infancy. Thomas retired to Market Rasen, where he died in 1704, leaving his residuary estate to his wife Ann and his son William (then however only six years old) as joint executors.[18]

These are the bare bones of William's known genealogy. It is difficult to put much flesh on them, but certain features of the foregoing account stand out. Most obvious is the very local nature of his background. His birthplace was probably only a mile or two from where his great-great-grandfather had farmed in the late sixteenth century, and less than four miles from Holton-le-Moor. So when he eventually became a landed proprietor it was in a district where he had deep roots.[19] Secondly, in the intervening generations the family had consistently retained Market Rasen as its market town. The only place mentioned in this family history that falls outside that market area is Wragby, a small market town in its own right some seven miles south of Market Rasen. Thomas's marriage to Ann Waltham

---

[16] Lincs. Archives, Lincoln diocesan records: Act Book V fo. 97v, Ad Acc 3/21–22, INV 85/392; Monson MS CXCIII, in private possession.
[17] Lincs. Archives, INV 160/154.
[18] LCC Wills 1704/1807; pedigrees compiled by T.G. and G.S. Dixon c.1906–40, in private possession.
[19] Alan Everitt noted that families that attained influence in a neighbourhood were often drawn from 'the indigenous families of the area in question' (*Landscape and Community*, 313).

of Wragby was still a local one, but it shows perhaps that the family was beginning to extend its range.[20]

Thirdly, these Dixons were not typical *peasants*.[21] William of Middle Rasen is perhaps the closest they got to joining the native peasantry of the district: Middle Rasen was a settlement of small farmers and small proprietors, and William's dual occupation was probably quite typical of it. But the Dixons did not put down roots in Middle Rasen or connect themselves by marriage with it. The two Robert Dixons of Owersby had been tenant farmers, holding land in a large and prominent grazing parish owned by one of north Lincolnshire's leading aristocratic proprietors. Thomas Dixon (d. 1704) was also a tenant farmer, but notably more mobile than his Owersby forebears, and it was perhaps his career that, despite his early death, provided the springboard for his son's rise into a more exalted stratum of the farming class.

There is, however, another factor in William Dixon's background that does not emerge from his direct ancestry, at least as far as it has been possible to trace it. Around 1700 there were a number of Dixons living and farming within a few miles of Market Rasen who could claim descent from a yeoman family of that name living in the early seventeenth century at Owmby-by-Spital, a village on the Lincoln Cliff some seven miles to the west of Market Rasen itself. In the first half of the eighteenth century one branch of this family became quite substantial landowners. Joseph Dixon of Buslingthorpe (1694–1751) acquired property in Middle Rasen and elsewhere, which descended to his eldest son Joseph, who was High Sheriff of the county in 1759 but died in 1763. This Joseph was succeeded by his younger brother Marmaduke, who on his death in 1780 left land in Middle Rasen to his cousin Marmaduke Tomline of Riby, near Caistor. Other branches of these Owmby Dixons farmed at Normanby-by-Spital, near Owmby, and a little farther south at Spridlington. William Dixon and his son Thomas were to become connected with both Normanby and Riby, which makes it all the more frustrating that no genealogical connection can be established between the these two Dixon clans. All one can say is that they most likely had a common ancestor in the late fifteenth or early

---

[20] W.G. Hoskins found that Leicestershire yeoman families tended to remain rooted in an area of six miles radius or less (*Leicestershire Yeoman Families and their Pedigrees* (1974), quoted in Phythian-Adams, *Re-thinking English Local History*, 33).

[21] To qualify as true peasants, according to Alan Macfarlane, *The Origins of English Individualism* (Oxford, 1978), the Dixons would have needed to become embedded as small family farmers in one community (such as Middle Rasen) and to have developed kinship ties within it.

sixteenth century, before the period of parish registers. There are two documentary clues, however, that suggest some contemporary awareness of a link. A collection of deeds surviving from the office of a Barton-on-Humber solicitor includes items relating both to the Holton Dixons and to the Owmby and Middle Rasen ones; and a lease of the Normanby rectory estate to Robert Dixon of Normanby in 1717 was reused as the cover of a Latin phrasebook for William's son Thomas in 1741.[22]

* * *

Ann Dixon, left a widow with a young son, soon married again. Her second husband was a Market Rasen neighbour, Sailbanks Broughton, himself a widower with three boys all younger than William Dixon. Broughton was a saddler, and the son of a Market Rasen innkeeper, but at some date between 1704 and 1720 he and his family moved to Holton-le-Moor. He died in 1731, his wife having presumably predeceased him. It was his son Benjamin – William Dixon's step-brother – who made the purchases in Holton from the Bestoes already described.[23]

Little, alas, is known of William's early life. There was a family tradition that he went to Market Rasen Grammar School – he was certainly better educated than his father – and that as a schoolboy he would sell back to his fellow-scholars the marbles that he had won at play, thus giving an early demonstration of his financial acumen.[24] Whether or not he ever lived at Holton with his step-family is unrecorded, but in the early 1720s he was set up as a tenant farmer at Claxby, a parish that shares a short length of boundary with Holton at the latter's south-east corner. Perhaps his mother had paid for the entry, or perhaps the capital was inherited by William at her death. In either case she was no doubt a crucial influence in his establishment as a young farmer. In 1723 he took a pauper child from Holton as a farm boy, and he served as churchwarden at Claxby the following year. He married in about 1726, and this was probably when he moved to the neighbouring parish of Normanby-on-the-Wolds, where his

---

[22] Lincs. Archives, Barton Parish Deeds, BPD 6–7; DIXON 1/A/1/14; Dixon of Owmby pedigree, in private possession.

[23] At this period the Broughtons were of a somewhat higher social position than the Dixons. The name Sailbanks had been brought into the family by a Rector of Wainfleet All Saints who died in 1622. Benjamin Broughton married Mary Raisbeck, daughter of the Rector of South Kelsey St Nicholas, and it was her marriage portion of £300 that helped him to buy the property in Holton (DIXON 1/A/4/1–2).

[24] Notes on family history by T.G. Dixon (DIXON 11/6/1).

children were born, and where he was churchwarden in 1727–8, 1731 and 1734.

William's bride was Rachel Drewry, of Adlingfleet in Yorkshire. Her father, Ellis Drewry, was a man of substance and very much a Yorkshireman. He was a freeman of York; he had married at Howden; and his son Roger was to spend his last years at Rufford Hall, near Harrogate. Apart from owning land and houses at Adlingfleet Ellis held land under Sir Michael Warton (who lived at Beverley), carried on business as a brewer, and probably also leased the Adlingfleet tithes. When he made his will in 1727 he was in 'an indifferent state of health', and he died the next year, leaving his daughter Rachel Dixon the family house, one or two other small pieces of property in Adlingfleet and £10 in cash. This did not exactly make Rachel an heiress, and may have represented no more than a delayed marriage portion. The rest of his property went to Roger, and the house was also to revert to him should Rachel have no heir.[25] However, it would seem that Ellis intended his son-in-law to succeed to the management of the Adlingfleet farm and his other local interests, a mark of confidence in him as well as an indication perhaps that Roger had no wish to take them on himself.

William Dixon had apparently entered a world at some remove from that in which he had been brought up. We do not know exactly how the connection was made, but in studying the Dixons and their milieu one gets used to the fact that important turning points in the history of the family generally had their explanatory antecedents, whether or not they left traces in the subsequent record. Adlingfleet, to start with, was not on the other side of the globe. It was only just the other side of the Lincolnshire boundary, and not much more than twenty miles as the crow flies from Claxby and Normanby. As an enterprising young grazier William may well have established links with the fertile marshlands near the mouths of the Trent and the Don, where his wife's parish lay. But there may possibly have been connections closer to home. Rachel's family had been at Adlingfleet since the early seventeenth century,[26] but at the same period there had been Drewrys or Drurys in the Market Rasen area, and a survey of Owersby in 1608 mentions Jenkin Drewrye's yard as well as Dickson's Close and Edward Dickson's yard.[27]

---

[25] Lincs. Archives, Foster Library, FL Garthorpe and district deeds.
[26] *Ibid.*, 1/20. They lived at Kirkgarth House, to the north of the churchyard.
[27] Monson MS VIII, in private possession. The Revd John Johnson of Cammeringham, who was related to the Owmby Dixons, left a legacy to his daughter Elizabeth Drewry in 1746 (BPD 1/6).

As far as one can tell William maintained his wife's Adlingfleet interests during her lifetime. There are references in his papers to the tithes of that parish as late as 1749–51.[28] (Mrs Dixon died in 1752.) But apparently there was never any question of the couple living permanently there. William continued to farm at Normanby, under the Markham family, until 1740.[29] And in the mid-1730s he took a large farm, of some 430 acres, under Lord Monson at Owersby, at an annual rent of £130.[30] This made him the third largest occupier in the parish, and provided a very useful complement to his wold land at Normanby. The year 1740, however, marked a change in his business. On giving up Normanby he and his wife and young son did move to Adlingfleet, but this was only a temporary arrangement. The following year, as we have seen, he made the purchase in Holton, where he intended to live and farm while still retaining the Owersby holding. The manor farm at Holton did not come with vacant possession, but he was able to secure a lease of Ewefield, in Holton parish, a good grazing farm with its own house, and he probably took up residence there at Lady Day 1742.[31] Ewefield and his Owersby farm were only about three miles apart, and could be worked together quite conveniently.

Now in middle age, William Dixon had built up a good grazing and farming business, and had become a landed proprietor. But, however significant the Holton purchase may appear in retrospect, at the time he may not have seen it as part of any long-term strategy. Even as an investment it had only limited potential. The house was not a good one, the farm was mainly on poor soil, the Moor was relatively unproductive, and the manorial rights probably did not amount to much by the early eighteenth century. Benjamin Broughton secured a clause in the sale of the lordship exempting his own property from manorial service, but this is no proof that such service had been regularly exacted: it is more likely that he was guarding against the revival, by a resident and more active lord, of rights that had lapsed under the Bestoes.[32] It is true that Holton, unlike most of the neighbouring parishes, was in divided ownership, and might therefore

---

[28] DIXON 21/3/1, 22/8/1 fo. 2v.

[29] Or so one may deduce from a rental of the estate (East Riding of Yorkshire Archives, Chichester-Constable Papers, DDCC/151/26).

[30] Monson MSS CLXXXV, CLXXXVI, in private possession; Lincs. Archives, Monson papers, MON 9/2B/10–11. William's farm, or part of it, lay near the South Beck, and family tradition places it at South Gulham, between South Owersby and the river Ancholme.

[31] DIXON 1/A/2/22.

[32] DIXON 1/A/1/12.

provide the occasional opportunity for William to expand and consolidate his property. But in 1741 the odds were against much land coming on the market in the near future: most of Holton's absentee landlords, including the owner of Ewefield, had put their properties into family settlements.

More immediately William had to make two major adjustments to his life. He had not only become a landowner in his native district but as resident lord of the manor of Holton had acquired a new set of local responsibilities. And he had also moved to a place that fell clearly into the market area of Caistor rather than that of Market Rasen. This may not have meant a sudden or dramatic change, since he was no doubt already familiar with Caistor's sheep fairs, and in any case Caistor was too small to meet all his commercial needs. But nevertheless there must have been a change of habits and shift of allegiance.

* * *

For William the 1740s were a period of consolidation for his grazing and farming business. The lease of the manor farm at Holton fell in in 1745, and it was no doubt then that he took it into his own hands. This brought his total holding to over 700 acres. Both his son and his grandson were to have farms of a similar size, and it may have been regarded as the maximum that could be comfortably managed without having to employ a bailiff. Apart from his operations at Holton and Owersby he would have had to make the occasional journey to look after his wife's interests at Adlingfleet. And there were parochial matters to attend to at Holton: he was constable in 1743 and churchwarden in 1746. But one may guess that he did not spend a great deal of time or money on visiting or entertaining. Neither he – apart from the step-family at Holton – nor his wife had close family connections locally. And the household at Ewefield would have been a small one. Jane, their first-born, had died in infancy, and Thomas, born in 1729, remained the only child. (By naming his son Thomas, William had continued the family tradition of alternating these two names in successive generations. The custom had begun in the seventeenth century and survived into the nineteenth.) By 1741 young Thomas was at school learning 'phrases taken out of Cornelius Nepos',[33] and, more usefully, acquiring a good clear hand and some ability with figures. By 1749 he was

---

[33] 'Thomas Dixon his phrase book 1741' (DIXON 1/A/1/14). The volume survives because it was used to make a copy of the Bestoe/Broughton conveyance in 1742. The cover, already mentioned, was made from a lease of the rectory of Normanby-by-Spital to Robert Dixon, of Normanby, yeoman in 1717 (cf. Lincoln Dean and Chapter records, Cj/13/3/10).

helping his father on the farms, and making memoranda in a pocket book that was to remain in use until the 1790s.[34]

In 1747, only six years after the Holton purchase, William laid out another sum in land. The property was some way from Holton, and was bought not for farming purposes but as an investment.[35] For about 50 acres of good land in Skidbrook, Theddlethorpe and Great Carlton, on the Lindsey Marsh east-north-east of Louth, he paid £1,250, or twenty-five years' purchase on the rental of £52 2s 6d. He does not seem to have needed a mortgage, so this considerable sum must have come mainly from business profits, although it is possible that part of it came from a progressive liquidation of the Adlingfleet assets. We do know that the purchase was intended to secure an income for Rachel should William predecease her, and in fact the conveyance is in the form of a rudimentary settlement, the first in the history of the Dixon family, though with no trustee to preserve the remainder.[36]

These pieces of marshland had been the property of John Smith, of Louth, a fuller and dyer, and were being sold by his daughters, one of whom was Mrs Cotes of Walesby. Her husband John was a substantial grazier, and Walesby was a parish only two miles south of Claxby and Normanby-le-Wold. It is likely, therefore, that Dixon had heard that this land was coming on to the market through his former neighbour Cotes. This, moreover, is not the only time that Cotes's name will appear in this chapter, and it seems legitimate to speculate that he had befriended the young and fatherless William Dixon in the early years of the latter's farming career. There may even have been an awareness of a family link: Cotes was connected to the Dixons of Normanby-by-Spital, where his own family had farmed since the early seventeenth century.

As it turned out William was to out-live his wife by many years. She died in 1752. In 1755 Thomas left home, and about a year after that William decided to move from Ewefield, an isolated and now lonely farmstead amidst its fields, into the village of Holton, where he took up residence in

---

[34] DIXON 22/8/1.

[35] Parts of the Marsh property were to be taken in hand by the Dixons from time to time, but it was never an essential adjunct to their farming business. For eighteenth-century Lincolnshire farmers who bought marsh land as an investment rather than for summer grazing see B.A. Holderness, 'The English land market in the eighteenth century: the case of Lincolnshire', *Economic History Review*, 2nd Series, 27 (1974), 570.

[36] DIXON 1/B/4/1–2; 4/1, fos 110, 133.

his house near the church.[37] He had no intention of retiring from farming, however, and retained both Ewefield and the Owersby holding. He had therefore to set up Thomas, who was ready to embark on his farming career, in a holding of his own. The farm that William found for him in 1753 was at West Firsby, on the Lincoln Cliff between Owmby-by-Spital and Spridlington. It was a good ten miles from Holton, and although that was a manageable ride it was a somewhat out of the way place. It was however a large grazing farm, of about 700 acres, capable of improvement, and it may have been drawn to William's notice by John Cotes, whose property at Normanby-by-Spital was only a couple of miles to the north. At first Thomas was the junior partner in the enterprise, but after two years he became solely responsible for its management, although his father remained the nominal lessee.[38] In May 1755 Thomas set up house at Firsby, and in August of that year he took his new bride there.[39] His marriage and subsequent career will be described in the following chapter.

* * *

During the 1760s, although past sixty himself, William showed no sign of wishing to retire from his farming and grazing business. Agriculture was doing better than it had for some time past, and he was probably enjoying a supplementary income from letting his rams. He also acted as commissioner for three parliamentary enclosures in north Lincolnshire, at Searby in 1763–5, Rothwell in 1765–7 and Barnoldby-le-Beck in 1769–71, probably earning himself a few hundred pounds in the process.[40] (These, incidentally, were all parishes that were later to have connections with the Dixon family.) His expenses, on the other hand, remained low. He no longer had a wife to provide for, and his son was now earning his own living. It is no wonder, therefore, that his wealth steadily increased. Like other well-to-do farmers he lent money on bond to trustworthy friends and neighbours. He had £200, for instance, out to Thomas Parker of Owersby, a loan that was transferred on Parker's death in 1769 to Michael and Whar-

---

[37] 1756 is the last year in which he appears in the parish account book as 'of Ewefield' (DIXON 17/4/15).

[38] Assignment of lease to William Dixon of Holton-le-Moor, gent., 10 July 1753 (DIXON 1/E/5/2).

[39] DIXON 4/1; 22/8/1.

[40] *Ex inf.* Mr Rex Russell. Commissioners were named in the relevant local Act of Parliament and had the responsibility of carrying it out. For sums paid to enclosure commissioners at a slightly later date, see Eleanor and Rex C. Russell, *Making New Landscapes in Lincolnshire: The Enclosure of Thirty-four Parishes in Mid Lindsey* (Lincoln, 1983), especially 116.

ton Rye, also of Owersby.[41] He invested £600 in the Ancholme Drainage and Navigation, a project from which his Owersby farm stood to benefit; and he also bought £100 worth of shares in the Bishop Bridge turnpike, the road that carried the produce of the Market Rasen area to the Trent and the Midlands. But for larger investments William's local choice was limited: indeed, apart from lending money on mortgage, which he does not seem to have done, his principal option was to extend his landed property, whenever opportunities arose that were both desirable and affordable.

In 1762 he made his first purchase of land for fifteen years, acquiring a farm of about 267 acres at Normanby-by-Spital. The land was leasehold of the Dean and Chapter of Lincoln, and he paid £1,400 for the unexpired portion of a twenty-one-year lease granted in 1760. The rental income – the property was an investment, and he had no intention of farming it himself – was low, only £76 a year in 1780, but the fines payable on the renewal of the lease, though large, were infrequent, and taking the fines into account he could expect an annual return of between four and five per cent.[42] The connection was once again the Cotes family, the farm having come up for sale on the death of John Cotes himself.[43] One of William's new tenants was Elizabeth Bell, a granddaughter of Christopher Dixon of Owmby: this Christopher may well have known William's father, since they had both been farming at Kingerby in the 1690s.

William spread the cost of the Normanby purchase over three years,[44] and he had scarcely settled the account when the opportunity arose for a much larger acquisition in the immediate neighbourhood of Holton. It was probably in 1766 that he paid about £3,400 for a leasehold moiety of the parish of Thornton-le-Moor.[45] This time the freehold belonged not to the Dean and Chapter of Lincoln but to the Bishop of Ely. The parish was held on leases for lives rather than for terms of years, and there were two lessees who each held an undivided moiety of the estate rather than a separate half of it. Whereas at Normanby the out-rent was low, the Chapter's income being mainly in fines, at Thornton the lessees were responsible for

---

[41] DIXON 4/1, fo. 121.
[42] Lincoln Dean and Chapter, Cj/13/1; DIXON 1/E/5/3.
[43] Cotes had been nominated as one of the enclosure commissioners for Keelby, but died before the award was made.
[44] DIXON 1/E/1/14.
[45] Cambridge University Library, Ely diocesan records, CC Bp 94511. The date must be between 1763, when the lease had last been renewed, and 1767, for which year a rental survives in the Dixon papers (DIXON 1/E/9/1). In 1766 Thomas Dixon lent his father £300, indicating a sudden need for money on William's part.

an annual rent to the Bishop of £200 gross, although by the late eighteenth century this had shrunk to about £140 net. This sum the lessees recouped in rents from their five farming tenants. This archaic and rigid system gave little scope for improving the parish while it remained undivided, and of course as a lessee William had acquired no specific amount of land that he could take in hand and farm himself. But being, unlike the other lessee, resident in the neighbourhood he could at least take on the role of agent and rent collector.

It was ironic that soon after making these acquisitions in Normanby and Thornton William should have been presented with a chance to buy land in Holton itself. The properties on offer represented the Willoughby, formerly Rothwell, estate in the parish. Stephen Rothwell had died in 1716, leaving his substantial north Lincolnshire estates, including about 550 acres in Holton, to his great-nephew Thomas Willoughby, younger son of the first Baron Middleton. Willoughby died in 1742, being succeeded by his son Henry, and it seemed likely that these estates would descend in the Willoughby family for many years to come. In 1752, however, Henry Willoughby disentailed them. He kept Thorganby, Croxby, Rothwell and Binbrook, but disposed of his land in Holton between 1769 and 1775.[46] William could not afford to buy all of it, but concentrated on the land east of the Rasen–Brigg road and thus immediately adjacent to his own estate. For the land west of the road a purchaser was found in Cornelius Stovin. Stovin lived at Hirst Priory, near Crowle, in the Isle of Axholme, but was connected by marriage with the Caistor area through a marriage alliance with the Turner family.

William's first purchase from Willoughby was Joseph Barkworth's farm in 1769. (For purchases of land in Holton by successive members of the Dixon family, see Map 2.) It lay mainly to the north of Manor Farm, on the other side of the sandy lane that led towards Nettleton and Caistor, and like Manor Farm it included one of the Maze fields. (The Maze was an area of meadow, formerly open but by this date divided into four closes: see the key to Map 2.) Its total size, however, was only about 58 acres, and its high price of £1,385 is accounted for by the fact that the sale included the warren.[47] For William this was a chance not to be missed. The farmland lay near his own, and was later to be treated as part of Manor Farm; and the right of free warren at last gave him full control over the manorial

---

[46] Nottingham University Library, Middleton Papers, Mi Da 67, 73; Mi E 17. Willoughby succeeded a cousin as fifth Baron Middleton in 1781.
[47] DIXON 1/A/3/1–2.

waste that he had owned since 1741–2. In 1775 he followed this with the purchase of William Jacklin's farm, which lay south of the village, between the Rasen road to the west and the Moor to the east. For these 77 acres he was constrained to pay the very high price of £1,100, a figure calculated on over thirty years' purchase at its then rental.[48] Finally, in 1777, he bought a cottage and close next to Barkworth's and later known as Noble's, but this was from Stovin rather than directly from Willoughby.[49]

The effect of all these acquisitions was to give Dixon control of the eastern side of the parish, except for two of the Maze closes, the village property belonging to the Broughtons, and a small close that formed part of the residue of the Bestoe estate. They brought his total estate in Holton to around 825 acres, although of this as much as 600 was moor or rabbit warren. His total estate in north Lincolnshire, however, both freehold and leasehold, now stood at nearly 1,750 acres. It had cost him well over £9,000, but by the end of his life his rental income from it was just over £450 (including a notional £60 for Holton Manor Farm), or about five per cent on his outlay.[50] Although he had at times been obliged to resort to short-term loans he had managed to avoid incurring mortgage debts, and in old age was enjoying a very comfortable net income, the fruits of a long life and, one suspects, a single-minded devotion to business. It is noteworthy that apart from Manor Farm and an acre or two of the Marsh property these purchases were not made with the intention of extending the acreage that he farmed himself. When he bought land as an investment he was prepared to assemble a portfolio that extended over the county from Normanby-by-Spital in the west to the coastal Marsh in the east. That said, when it came to his home territory of Holton he could brace himself if necessary to pay over the odds to consolidate his estate and prevent adjacent land from falling into possibly undesirable hands.

* * *

Prosperous though he became, there is no evidence that William Dixon changed his habits of life to correspond with his increasing wealth. In old age he continued to live quietly in his very modest house at Holton, with

---

[48] DIXON 1/A/3/3–4. According to Holderness ('English land market', 576), the median price of land in Lincolnshire was twenty-nine years' purchase between 1761 and 1775, but fell to twenty-five years between 1776 and 1790.
[49] DIXON 1/A/3/5.
[50] DIXON 4/1, fo. 110.

probably only a housekeeper to look after him.[51] In the agreement for the Holton purchase in 1741 he is described as 'grazier'. Thereafter, as a man of property and of high standing in his community, he was consistently credited with the title of 'gentleman' to which he was fully entitled. But when he came to make his will in 1780 he chose to revert to 'grazier'.[52]

The previous year, at the age of eighty-two, he had given up the Owersby farm. The incoming tenant was William Thorpe, a well-known north Lincolnshire grazier, who took the flock and livestock at a substantial valuation.[53] Even then the old man continued farming at Holton, but his young grandson and namesake, William, came to help him and to provide some companionship in his final years.[54] William senior seems not for some years to have seen a great deal of his son Thomas, who was leading a busy life some miles away, but between him and his eldest grandson William there seems to have been a stronger affinity. Having made his will, a very simple one, in November 1780 he allowed young William to take over the reins at Holton.[55] He survived the winter of 1780 but died, aged eighty-four, in July 1781.

---

[51] William left £10 in his will to a Sarah Longbottom, possibly his housekeeper. He also left one guinea each to his 'servants', but these may have been farm servants. (DIXON 2/1/1.)

[52] See also Penelope J. Corfield, 'The Rivals: landed and other gentlemen', in Negley Harte and Roland Quinault (eds), *Land and Society in Britain 1700–1914* (Manchester, 1996), 6–21; R.J. Morris, *Men, Women and Property*, 82.

[53] DIXON 1/E/5/4. The names of the valuers – David Young of Thornton-le-Moor and John Walesby of Swinhope for Dixon and Benjamin Codd of Glentworth and Walter Nunwick of Saxby for Thorpe – indicate that an important flock was changing hands.

[54] There is no direct evidence for this, but it fits with the surviving evidence, and accords with family tradition.

[55] DIXON 4/1, fo. 110.

3

THE TENANT FARMER: THOMAS DIXON (1729–1798)

Thomas Dixon, as we have seen in Chapter 2, was the only surviving child of William and Rachel. By 1753, when William took the West Firsby farm for him, he had already gained a few years' farming experience. For the first two years at Firsby he was his father's manager, as already mentioned, but from 1755 he was the tenant of the 700-acre holding in all but name. A painstaking and methodical young man, he began to keep the detailed accounts that remained a lifelong habit, and from which one can construct a more substantial narrative of his affairs than is possible for his father.[1]

West Firsby was primarily a grazing farm. It could support a flock of about seven hundred sheep – approximately one to the acre. On the lower part of the farm there were pasture closes and 92 acres of meadow, but to the west the land rose through sheep walks to an area of furze and bracken near Ermine Street, the old Roman road that led south towards Lincoln.[2] The sheep were in all probability the small and hardy variety suited to the uplands of the county, not the large longwools that grazed in such numbers on the lusher pastures of the Lindsey Marsh.[3] In fact the farm could not support the whole flock throughout the year, and turnip fields were hired from neighbours such as 'Mr Maurice' and 'Mr Carr'.[4] As was usual, the flock was maintained and improved by hiring rams from elsewhere rather than through inbreeding. In November 1756 Thomas borrowed his father's 'great tup'.[5] In addition to the flock he had a small herd of cattle: he sent some beasts to London in 1755, and there was also some butter-making.[6]

---

[1] DIXON 4/1–2 *passim*. No account book for William survives, but from Thomas's one can deduce something of William's grazing methods around 1750.
[2] DIXON 22/8/1, fos 3v, 9.
[3] J.A. Perkins, *Sheep Farming in Eighteenth and Nineteenth Century Lincolnshire* (Sleaford, 1977), 6ff.
[4] DIXON 4/1, fo. 2. Probably Morris of Spridlington and Carr of Normanby-by-Spital.
[5] DIXON 4/1, fo. 3.
[6] DIXON 4/1, fo. 137.

But little corn was grown, apart from some barley apparently for home consumption.

A pastoral farm was said to make two rents, and this seems to have been the case at Firsby. The rent was £200 a year, and Thomas's incomings averaged around £400, with a third of his gross income coming from sales of wool (to a Yorkshire merchant no doubt inherited from his father), another third from the sale of sheep, and the remainder from the sale of beasts and horses and other small transactions. The farm was therefore one from which a good living could be made, provided that one avoided the perennial hazards of bad seasons, sheep or cattle disease and slumps in the wool or livestock markets. But Firsby also had drawbacks. The house, although recently repaired after a fire, was probably not much more than a ground-keeper's cottage.[7] And the place was an isolated one: Spridlington and Owmby-by-Spital, the nearest villages of any size, were over a mile distant, and Market Rasen, the nearest market town, was considerably farther away over poor roads.

These disadvantages would have been particularly keenly felt by Martha Walkden, the young bride who came to join Thomas at the farm in 1755. She came from Great Limber, on the Wolds beyond Caistor, and West Firsby must have seemed to her a very distant and uncongenial spot. Not only would she have missed her immediate family, but she had been used to a local society more densely and variously textured than that in which Thomas Dixon had been brought up. Her father Thomas Walkden was both a farmer and a clergyman. As a farmer he occupied a substantial holding at Great Limber, on the Pelham family's Brocklesby estate, a farm to which he had succeeded by marrying the daughter of the previous tenant.[8] As a clergyman he had held various livings or curacies on a temporary basis in order to facilitate clerical arrangements by his landlord: he was, in other words, a useful cog in the machinery of patronage on the Brocklesby estate.[9] At the time of his daughter's marriage he was non-resident vicar of Cadney, near Brigg, and curate of Keelby, a populous parish bordering Brocklesby and Limber to the east. Many parsons, then and later, farmed their own glebe, but Walkden's dual occupation was unusual even for a period that took a relaxed view of clerical residence and sources of income.

---

[7] DIXON 1/E/5/1–2.
[8] Lincolnshire Archives YARB 5/2/1/2: Brocklesby estate rental 1765.
[9] Lincoln Diocesan Records, Register 38; SPE 1; *Keelby, Parish and People 1765–1831* (Keelby, 1986), 34–7; information from Mrs Dinah Tyszka.

He was thus a busy man, but he was not a rich one, and his means certainly could not match those of William Dixon. He was unable to make a settlement on his daughter at the time of her marriage, although he was later to leave her £600 in his will. But he had connections, and one connection in particular was no doubt discussed with his daughter's potential father-in-law. It related, however, not to the Brocklesby but to the neighbouring Riby estate. Riby lay on the dip slope of the Wolds, adjoining Limber to the west and Keelby to the north. The parish was in single ownership, and Walkden was sole trustee for its squire, Marmaduke Tomline, who was still under age in 1755. The parish was a fertile one, divided into four good farms of a type also common on the Brocklesby estate,[10] and Walkden was in a position to put Thomas Dixon in the way of the next farm to become vacant. Martha could be confident that West Firsby was not a life sentence, and in fact the couple were able to move to Riby three years later.

It was thus fortunate for Thomas that he had met Martha. We do not know how they had come together, but as usual in the Dixon saga there was a hinterland of family and local connection. Marmaduke Tomline (1736–1803) was undoubtedly a gentleman. He was to inherit, when he came of age at twenty-three, a good estate of about 3,000 acres, a handsome house in its own park, and a position among the lesser gentry of the county. But his family was not an ancient one. In the early seventeenth century his ancestor in the male line had been farming at Cotness, in the West Riding, not far from Adlingfleet. Riby had been purchased in 1680, but the Tomlines had not lost touch with the farming society from whose ranks they had emerged. William Tomline (1690–1743) married Elizabeth Cary, whose brother Thomas was the leading Monson tenant at Owersby, and thus a near neighbour of William Dixon. And William Tomline's sister Laetitia had married Joseph Dixon of Buslingthorpe, whom we have already encountered as a representative of the 'other' Dixon family.[11] It was Joseph Dixon who became the senior trustee for the young Marmaduke Tomline, only seven when his father died, and Walkden, the other trustee, who was left as sole trustee when Joseph Dixon himself died in 1751.[12] It is even possible that William Dixon was approached, but declined, to become a trustee at this point. What is certain is that when Thomas Dixon moved to Riby it was not just as a tenant but also as a kind

---

[10] The four leading tenants in 1782 were Thomas Dixon, David Winship, William Torr and Thomas Skipworth (in succession to George Swallow) (Lincs. Archives, Lindsey Quarter Sessions, land tax return 1782; Lind Dep 29/5; Riby Parish 7/1).
[11] See above, Chapter 2.
[12] Lincs. Archives, Pretyman-Tomline deposit, 2 PT 2/18, 3/8–13.

of adviser or minder for Marmaduke. One suspects that the latter never had a great head for business.[13]

In 1758 Jonathan Turner's farm in Riby became vacant, and Thomas Dixon, with his family and his flock, trekked over from Firsby to begin his new life. The house into which he moved was almost certainly the one now known as Church Farm. A handsome brick house, originally thatched, it looks as though it dates from Marmaduke Tomline's minority, and may even have been built for the Dixons as its first occupants.[14] Turner's holding was enlarged for Dixon to 770 acres, although 70 of them were an area of furze at the top of the farm known as Cottagers' Wold, from which the inhabitants of the parish supplied themselves with fuel. The remainder of the holding was similar to Firsby, but it contained fields in different parts of the parish rather than lying within a ring fence.[15] This made it less easy to work, but it ensured that its occupant benefited from a range of soils. There was good meadow at the north-eastern end of the parish, where the chalk dipped under a layer of gravel, and higher up was chalk land capable of producing good crops of corn, roots and seeds. This was where Riby scored over Firsby: it could support a large flock, but it could also grow cash crops of wheat and barley, the latter mostly of malting quality. Thomas could also grow turnips, thus reducing his dependence on other farmers for winter feed. During the 1760s he converted parts of his sheep walks to arable, necessitating the employment of more labour and, in 1767, the erection of new farm buildings.[16] He continued to hire rams, but also began to use his own tups and even to hire them out to other graziers. He never became a leading breeder, but he seems to have moved with the times, gradually shifting from the specialised grazing of his father's day towards the mixed husbandry, based on sheep, barley and turnips, characteristic of the wold farming of the next century.

As far as Thomas's commercial dealings were concerned, the move from Firsby to Riby meant a transfer from the Market Rasen district to that of Caistor, rather as his father's move to Holton had done two decades

---

[13] He was never made a magistrate. At the end of his life he quixotically left his estate to Bishop Pretyman, whom he barely knew.

[14] The house was rethatched in 1774 (DIXON 4/2, p. 142), which would fit with a building date some 20–25 years earlier. It is often difficult to be certain where exactly farmers lived, at a period when farms were not always named and did not always lie within a ring fence.

[15] This arrangement of the holding probably dated back to the immediately post-enclosure period. It lasted until the early nineteenth century.

[16] DIXON 4/1, fo. 119. The large and handsome barn at Church Farm was a later improvement. It bears the date 1782 and the initials 'MT' (for Marmaduke Tomline).

previously. He could now meet fellow-farmers at the Caistor weekly markets, deal with the Caistor butchers and make use of the malt kiln in the town. But Caistor by no means met all his needs. From time to time he conducted business at Brigg or Market Rasen, attended the great horse fairs at Horncastle, and hired farm and domestic servants at the Barton or Keelby statutes; and Grimsby, with its coastal links to Hull and London, was only a little farther away than Caistor. Equally important, however, were the transactions that took place privately rather than in the public arena. He continued to sell his wool to West Riding merchants; some of his major sales of livestock took place outside the fairs; and there were numerous dealings of a very local nature – the sale of small quantities of corn to neighbours or workpeople, for instance, or the hire of rams or turnip fields. Some of these transactions did not involve the exchange of money: they were payments in kind or favours given or returned.[17]

The names of the neighbours with whom he most frequently did business may reveal something about the nature of his local network. Among his fellow-tenants at Riby it was the Winships and the Torrs with whom he had the most regular contact: Thomas Skipworth – whose family was to become so closely associated with the Dixons in the next generation – appears less often in Thomas's accounts. Among the Brocklesby tenants it was naturally his father-in-law Walkden who features most frequently until his death in 1773, and then Walkden's son Richard, who took over his father's farm. Other Pelham tenants, such as Thomas Nicholson of Keelby, appear from time to time, but as a group they are not as significant as the farmers and other inhabitants of lowland parishes outside the Brocklesby estate such as Aylesby, Laceby and Stallingborough. At the same time Thomas by no means neglected his links with the Holton neighbourhood. With his father, of course, he had regular dealings, borrowing his father's tup, for instance, and on at least one occasion lending him money.[18] Caistor was a convenient halfway meeting-place at which goods or money could be handed over. During the 1760s he also maintained connections with his father's fellow-tenants at Owersby, hiring tups or renting supplementary pasture from them and occasionally negotiating loans with them. In the early 1760s 'Mr Cary of Grasby', presumably one of the successful Owersby Carys who had bought land at Grasby, near Caistor, was renting land at Riby.[19]

---

[17] DIXON 4/1–2 *passim*.
[18] DIXON 4/1, fo. 121.
[19] Riby overseers' accounts in DIXON 4/1.

* * *

The 1760s and 1770s were busy years for Thomas Dixon – and for his wife. She had given birth to her first two children during the Firsby years, William (in fact born at Limber, where his mother had retreated for the event) and Rachel. At Riby there followed Thomas (1759), Richard (1760), Martha (1762), Mary (1764), Jane (1766), Marmaduke (1768), and Ann (1772). All except Marmaduke reached adulthood, although Martha died at the age of twenty-one. Thomas thus had eight children to educate, and eventually three boys to establish in business and four girls to provide for. At times the Riby household must have been crowded, although each successive baby was put out to nurse, and did not return home for at least a year. (Nancy, the youngest, stayed with Mrs Ranby, the shepherd's wife, until she was nearly two.) When the girls were old enough they were sent to local academies, where they acquired the standard middle-class accomplishments of the day. With the boys, however, decisions had to be made that would determine their future careers. Willy went in August 1765, as a nine-year-old, to the free school at Laceby,[20] and in 1772 to Brigg Grammar School, where he boarded with the usher, Christopher Cave. This was sufficient education for the son who was destined to follow his father into the farming business. Both Tommy and Richard, however, were educated for the Church. They followed William to Laceby in 1768 and then to Brigg in 1773. But at Brigg they were boarded not with Cave but with the Master, the Revd John Skelton, who prepared them for Oxford.[21] Tommy matriculated at University College in 1778 and Richard at Corpus Christi in 1779.

As a teenager William was given jobs around the Riby farm, beginning in 1775,[22] but by 1779, when he was twenty-three, it was time for him to acquire further experience away from home. It was agreed that he should go over to help his grandfather at Holton, where he could learn how to manage Ewefield as a grazing farm, how to supervise the warren, and how to administer the tenanted properties. Two years later William senior died, leaving £20 to his step-brother Benjamin Broughton, £100 to Benjamin's daughter, later Mrs Bett, but all his landed property to his son Thomas, who was sole executor. Thomas and his wife were too firmly settled at Riby to contemplate moving to Holton. Instead it was decided – perhaps

---

[20] For his Latin phrase book 1770, see DIXON 22/8/2.
[21] DIXON 4/2, fo. 50; F. Henthorn, *The History of Brigg Grammar School* (Brigg, 1959), 47.
[22] DIXON 4/1, fo. 17.

this had already been anticipated in 1779 – that William would remain at Holton to look after the family interests there. To provide him with a chance to earn his own living his father also let him Hall Farm, Taylor's Farm and the warren at a rent of £80 a year.[23] This was a low figure, but even so it was hardly a very generous gesture on Thomas's part. Ewefield would have been an obvious supplement to William's holdings, but at this point it went out of Dixon occupation.

The following year William got married. His bride was Amelia Margaretta Parkinson of Healing, a parish not far from Riby. Her father was the Revd John Parkinson, a well-to-do clergyman and landowner (of whose family we shall say more later in this chapter), and he gave her £1,000 on her marriage, with the promise of another £500 on his death, this latter sum being 'left entirely to his honour'.[24] In return Thomas settled a 'trostment' on her consisting of the marshland (previously used as a provision for his mother) and Robert Hall's farm, late Jacklin's, in Holton.[25] The two properties were reckoned to produce £100 a year, and Thomas also agreed to provide a house at Holton for the couple. The new Mrs Dixon would expect no less, and in any case it was time that the estate had a respectable mansion house to go with it. This house, of which more later, was erected between 1783 and 1785, at a cost of £500. But Thomas did not make a gift of it. Instead he saddled William with an annual charge of £25 representing the interest on the capital sum expended on it.

This arrangement left Thomas free to continue his life at Riby much as before. He neither contracted nor expanded his farming operations, but seems to have begun to delegate more of the routine work to his shepherd, and to have simplified his corn and sheep sales. There was perhaps more reliance than before on the Lincoln spring and Caistor autumn fairs, and he sometimes used his son William or the Nettleton stockjobbers Robert and William Saunderson to act for him. He wrote to his son William in April 1794 as follows:[26]

---

[23] DIXON 4/1, fo. 91.
[24] DIXON 4/1, fo. 104. See also Appendix 2, Table 2.
[25] A settlement was drawn up with Marmaduke Tomline and Richard Walkden as trustees (DIXON 2/2/1–2).
[26] DIXON 7/5/19.

Riby, 18 April 1794

Dear Son

I have been but indifferent this week but am rather better. I have not seen anything of Mr Saunderson this week. You will take care to send a drover to Riby to help Lancaster with Mr Tomline and my own shearlings to Holton, wou'd have them set off from Riby abt 8 o'clock on Monday morning. I shall take care to send the culls to Castor to take the fortnight market, and shall wish you to sell them, as you will form an opinion on their value by seeing Lincoln fair.

You will also tell the Thornton tenants abt paying their rents, which I suppose will be on that day.

You will also tell Saunderson I shall expect to see him on Monday or Tuesday about my beasts, otherwise I must sell them to some other person.

I wish you wou'd desire Mr Kelk junr to buy me a new pack sheet at Brigg. I shall want to put my locks in it when ready. I cou'd like to have it very soon.

I am your affectionate father

T. Dixon

PS. Remember me to your four sisters.

The impression this letter gives is of a somewhat inflexible and demanding old man, expecting his son to run errands for him despite the fact that the latter had by this date many commitments of his own.

Just as he made few changes in his farming routine in his later years, so Thomas neither expanded nor contracted the portfolio of landed property that he had inherited from his father. The one possible exception is a farm at Keelby, worth £78 a year, in the purchase of which he was involved in 1778.[27] But this appears to have been acquired for the Walkden family, and passed to Martha Dixon only when her brother Richard Walkden died in 1793. The value of the Normanby estate increased considerably in the 1790s, following the enclosure of the parish in 1795, but Thomas does not appear to have initiated this development, and to offset against the gain was the sum of £80 that he had to pay as his share of the enclosure expenses, plus another £245 to renew the lease.[28] Similarly the Thornton estate was divided in 1797, giving Thomas the northern half of the parish.

---

[27] The farm cost £1,675 (DIXON 4/1, fo. 107).
[28] DIXON 4/1, fo. 86.

This consisted mainly of two farms, Beasthorpe and Gravel Hill, and they were to play an important part in the farming activities of the Dixon family after Thomas's death. But again the immediate effect was to put Thomas to expense, both in connection with the division and with the erection of a house and buildings at Beasthorpe. William oversaw the work for him and dealt with the bills.[29]

In 1796–7 the elderly Thomas reckoned his total annual income at £840 3s. Of this the major part, £552, came from landed property, and comprised £200 from Thornton, £172 from Normanby, £105 from Holton and £75 from Mrs Dixon's Keelby farm. Another £97 5s came from interest on loans, including £1,000 to Marmaduke Tomline, £500 to his son William Dixon (representing the cost of the house at Holton) and another £500 to William's brother-in-law Robert Parkinson of Barnoldby-le-Beck. The Ancholme Navigation and Bishop Bridge turnpike securities were producing £34 16s a year. And he reckoned the 'profit of the farm' at £156 – a surprisingly low figure, even if it was net of all household expenses.[30] (Perhaps it was not a typical year: farming years seldom were.) Altogether the couple at Riby were very comfortably off, but their income was not enormous considering the size of Thomas's inheritance and the long years during which he had occupied a very good farm at a very reasonable rent.[31] His children had undoubtedly been the principal drain on his resources.

Or rather two of his children in particular. In the matter of William's first farm and his marriage Thomas had got off quite lightly. Three of the four girls were married in his lifetime – as will be described shortly – but he had been able to promise money at his death rather than pay their portions in full at the time of the nuptial contract. But the expense of Tommy and Richard had not come to an end when they left college. Tommy was the first to be settled. Thomas Winship, the rector of Laceby (and no doubt a relative of the Riby Winships) died in 1783, and his widow presented Tommy to the vacant living. In 1787 Tommy married, and Thomas Dixon was able to buy the right of next presentation to Laceby from Mrs Winship for his son as his contribution to the future of the family. It was a good match. Charlotte Woolmer was a well-educated lady with good connections: through her mother she was related to the Fields, the leading resident family in Laceby, and she brought £1,800, half of it however expectant on

---

[29] DIXON 4/1, fo. 45; 4/3, fo. 210v. Nearly £300 was laid out between March 1797 and October 1798.
[30] DIXON 4/1, fo. 90.
[31] He paid around £260 a year, rising to £300 only in 1798.

the death of her mother.[32] In 1792 Tommy received a pleasant supplement to his income when Pelham presented him to the vicarage of Eyeworth in Bedfordshire, where the Brocklesby family had an outlying property.[33]

In 1784, meanwhile, Thomas had acquired the living of Claxby with Normanby for his second clerical son Richard. The two parishes were familiar Dixon territory. They lay only three miles south of Holton, and Richard's grandfather had been churchwarden of Normanby, which was annexed to Claxby for ecclesiastical purposes. It was a good living, with the glebe and tithes let to Charles Pelham of Brocklesby, the owner of the parishes, for £216 a year. The sitting incumbent agreed to vacate the living in favour of Richard at May Day 1785, but such a proceeding, perfectly usual in farm sales or tenancies, was simoniacal, that is, illegal, when it came to Church livings. The incumbent had to continue as nominal rector for his lifetime, and meanwhile Thomas allowed Richard £40 a year as curate. He set up a bachelor establishment at Claxby in 1786, with contributions of linen and table silver from Riby.[34]

Martha Dixon, the daughter of a clergyman, now had two of her three sons settled as beneficed clergymen, and her other son had married a clergyman's daughter. She and her husband Thomas were conscientious Church people themselves – it was later recorded that Riby had had psalm singing three times a week – and they had helped to establish two parsons, one near Riby and the other near Holton, who could not only carry out their parochial duties but make themselves available to serve other nearby parishes that lacked resident incumbents.[35] But then things went wrong. Both Tommy and Richard plunged into debt, Tommy to the tune of over £4,000 and Richard, who had put his name to some of Tommy's bills, to a smaller extent. They had to make a joint settlement of their affairs, as part of which Thomas made over the Claxby advowson to Richard, who then assigned it to trustees for the benefit of the creditors. It was under these ignominious circumstances that Richard finally became rector of Claxby when the old rector died in 1794.[36]

---

[32] DIXON 4/1, fo. 84. Charlotte's father, Joseph Woolmer of Barton, had been connected with the Nelthorpes, and her mother was a Lely of Grantham, so, assuming Lincolnshire represented the bounds of the cosmos, she was quite a cosmopolitan young lady.

[33] DIXON 4/1, fos 44, 91; 21/3/3/5.

[34] Lincs. Archives, Taylor, Glover and Hill deposit, 2 TGH 3/A/4/2; YARB 5/2/1/5; DIXON 4/1, fo. 90 enc. The Revd Richard Dixon later married Frances Watson, sister of one of the Cato Street conspirators (Dixon pedigree book, in private possession).

[35] DIXON 7/1/7: William Dixon's notebook, 4 February 1809.

[36] DIXON 19/2/1; 3 DIXON 5/11/4; BPD 6/12; Bishop's Register 39. Oxford University may not have been altogether good for them as far as habits of economy were concerned.

* * *

Given Thomas Dixon's standing as a substantial farmer with respected family connections, and as a leading resident of his parish, it was inevitable that the Dixons, from their early days at Riby, should occupy an assured position in the middle-class society of their district. But they appear to have given themselves no airs and cultivated few graces. They kept a chaise, and that meant keeping a mare to pull it and a groom or 'waiting man' to look after it. There does not seem to have been anything very smart about it, however, and it was never upgraded to a four-wheeled carriage. The groom hired in 1786 was to have 'an old coat, waistcoat and pair of breeches, a great coat and a hat with a button and loop'; and on at least one occasion the chaise was requisitioned to collect a small consignment of wheat.[37] Indoors Mrs Dixon managed with only two maid-servants, presumably relying on her daughters when they were at home to help her run the household. There are also indications that one or two farm servants lived in. William Cortis came to his place at May Day 1785 but got married in grass-mowing time and ran out of the house in the night, leaving the doors open.[38]

Cortis was not the only troublesome servant: in fact the Dixons had regular difficulties with their employees. In 1782, to mark the improvement in their circumstances, they paid a little more for a superior maid, only to find that she thought the place beneath her. William East, engaged as a groom in 1789, killed a mare belonging to 'daughter Dixon of Laceby', and ran off after four months. He 'ought to have been punished', opined his ex-employer, but he was interceded for by Richard Goulton of Croxby. The year beginning May Day 1794 was a particularly grim one indoors, when they hired a 'deceitful, lying, impudent jade' at £5 and 'a lying, impudent, saucy baggage' at £4 10s.[39]

Mrs Dixon would have used the chaise for going into Caistor, and for calling on family members at Limber and Laceby. Who else formed her regular circle of acquaintance we do not know, but it is likely that it featured farming families with clerical connections, and parsons themselves such as Mr Holiwell of Irby (who also held the Riby living), rather than the more bucolic and hearty farmers of the hunting and coursing type. Certainly on Mrs Dixons's list would have been her son William's in-laws,

---

[37] DIXON 4/1, fo. 28; 4/2, p. 136.
[38] DIXON 4/1, fo. 25.
[39] DIXON 4/1, fos 25–36. Goulton, however, was prepared to invoke the full force of the law to protect his rabbits in Croxby Warren.

the Parkinsons. They had produced both farmers and clergymen for generations, the farmers tending to be called Robert and the parsons John. Originating in the Scunthorpe area in the early seventeenth century, they were farming by the mid-eighteenth at Ravendale, on the Wolds not far from Caistor, and they had also acquired land at Healing. Robert Parkinson of Ravendale (1690–1740) married the daughter of one of the farming families at Riby, and after his death she married Thomas Cary, Marmaduke Tomline's father-in-law.[40] Robert Parkinson left his Healing property to his second son John, the rector of that parish, and it was one of John's children, Amelia Margaretta, who married William Dixon. Thomas Dixon had dealings with John's brother, Robert Parkinson of Barnoldby, and he would also have known the Revd John Parkinson of the Ravendale family, cousin of John and Robert, who became rector of Brocklesby in 1785 and who was a frequent guest at Brocklesby Hall.[41]

If William's marriage showed the family extending its range socially, the marriages of three of his sisters showed it consolidating its position over a wider area of middle-class Lincolnshire. They did not marry clergymen, but two married merchants and one an enterprising young farmer. (The fourth, Mary, did not marry until some years after her father's death: her husband was a Yorkshire doctor.) In 1784 Rachel married William Etherington, from a Gainsborough family of merchants and shipowners. How they met is not recorded, and the Middlemarch-like society of Gainsborough must have been rather unlike the rural neighbourhood that she was used to. In 1793 Jane followed her elder sister in marrying into the commercial aristocracy of a north Lincolnshire market town, but this time the town was Brigg, and the business connection with the Dixon family more obvious: Jane's husband John Kelk junior was the son of a leading Brigg merchant who supplied both her father and her brother William with seeds.

Ann, the Dixons' youngest and perhaps most favourite daughter, married at the end of her father's life, in 1798. Her husband Richard Roadley belonged to a family that owned and farmed land at Messingham, a few miles north of Gainsborough, with a smaller property at Hibaldstow, some three miles south of Brigg. Like the Dixons the Roadleys had come of local farming stock, but by the late eighteenth century they were not only prosperous (the Messingham estate ran to some 550 acres) but also more

---

[40] Robert Parkinson (d. 1740) was buried at Riby.
[41] For John Parkinson's diary notes 1779–1808, see DIXON 16/1. He later succeeded his brother in the Ravendale and Scunthorpe properties.

sophisticated than the Dixons.[42] They were not a notably clerical family, but in 1794 Richard's elder sister Charlotte had married the Revd William Jackson of Burton-by-Lincoln, who later, as we shall see, became rector of Nettleton, between Holton-le-Moor and Caistor.[43] Of all the marriages of Thomas's children, that of Ann to Richard Roadley would turn out to have the most significance for the history of Holton in the nineteenth century.

As for the wider contacts of the Dixon family in Thomas's time, there is no evidence that he ever got as far as London. The only clue that he travelled beyond the borders of the county comes from a little book of medical recipes, including one for the gravel collected at Buxton in July 1793.[44] And that visit was no doubt undertaken for health rather than for social reasons.

* * *

It was inevitable that Thomas should play a part in the affairs of Riby parish. In addition to the unofficial tasks that he took on for Tomline, including the management of a small sheep farm that his landlord kept in hand, he was also called on to act as overseer of the poor, and was still holding that office in the early and mid-1780s.[45] But by then his talents were in wider demand. He was appointed a tax commissioner in 1769, and from 1779 to 1786–7 he acted as agent for the Humberstone and Somerby estates of the Mackenzie family.[46] Then, in 1787, he became a county magistrate. This was unprecedented in his family, and not usually an office that was given to a working farmer. But there was a concern at the time about the shortage of resident magistrates in the county, and Charles Anderson Pelham, who put forward Dixon's name to the Lord Lieutenant, was particularly anxious

---

[42] Information from Miss Joan Gibbons. And compare the Roadley monuments in Riby and Searby churches with their plainer Dixon counterparts at Holton. For the Roadley family, see also Appendix 2, Table 3.

[43] William Jackson was known as 'newsy Billy Jackson' on account of his love of gossip. He was first cousin once removed of Bishop Jackson of Lincoln.

[44] Thomas's receipt book (DIXON 22/8/1) contained animal as well as human remedies, and was still in use in the 1860s when the Holton bailiff noted down a remedy for the cattle plague. Earlier entries include one collected from William Thorpe for the treatment of rheumatism and another from Marmaduke Tomline for keeping the body open.

[45] Thomas appears to have been rather at Tomline's beck and call. A letter from him in December 1783, addressed 'Sir', sends him money for 'our poor', and concludes 'If at leisure let me have your company about 4 o'clock' (DIXON 4/1, fo. 92 enc.).

[46] DIXON 4/2, p. 14; 8/5. Dixon had known the family for some time. Humberstone was near Grimsby, and Somerby near Gainsborough. Mary Humberston, the heiress of these estates, had married a Scot called Mackenzie, a descendant of the Earls of Seaforth.

that his own district should be adequately covered.[47] No local objection was raised by members of the existing Lindsey Bench, and in September 1787 Thomas was sworn in at the Gainsborough sessions.[48]

At that date the Lindsey sessions alternated between Gainsborough and Caistor. Dixon apparently found it too difficult to get over to Gainsborough after that initial meeting. He did attend some of the Caistor meetings, but in 1791 it was decided to centralise the sessions at Kirton-in-Lindsey, and he never attended there.[49] Although nearer than Gainsborough it was an awkward distance from Riby, and perhaps he also felt that the Kirton Bench was somewhat out of his social sphere. He could still be a useful magistrate, however, by staying at home and letting cases come to him.[50] For his neighbours his reluctance to leave home was a positive virtue, since he could deal summarily with minor matters brought to him there, and in more serious cases he could commit thieves and burglars to Quarter Sessions or even to the Assizes at Lincoln. Especially useful was his willingness to deal firmly with absconding farm or domestic servants: among those neighbours who most frequently prosecuted their workpeople were Joseph Johnson of Irby, Jonathan Winship of Riby and William Richardson, who succeeded to Richard Walkden's farm at Limber.[51] One may, however, question whether his very local approach to his judicial duties, and his close identification with the employers' side in contractual disputes, made him the ideally detached and impartial magistrate.[52]

---

[47] Lindsey Quarter Sessions, LQS/E/Justice of the Peace correspondence 1786–7. It is not clear how well Pelham knew Dixon, although he would have known *of* him. Possibly the Revd John Parkinson put in a word for him.

[48] He took advantage of the trip to Gainsborough to sell some sheep to Richardson of Willoughton (DIXON 4/1, fo. 102).

[49] James Green Dixon junior told T.G. Dixon in 1910 that Thomas, 'owing to a family characteristic, was usually late for meetings', and missed the important meeting at Caistor in 1789 that decided on the move to Kirton (DIXON 11/6/1). But it is doubtful whether his presence would have made any difference.

[50] B.J. Davey (ed.), *The Justice Books of Thomas Dixon of Riby 1787–1798*, in *The Country Justice and the Case of the Blackamoor's Head: The Practice of the Law in Lincolnshire 1787–1838* (Lincoln Record Society 102, 2012).

[51] DIXON 8/1–4; Davey, *Justice Books*.

[52] There was also the problem, for working farmers, of leaving their holdings unattended or of having to make arrangements to cover their absences when away on justice business. (For a working farmer and J.P. in south Norfolk in the 1790s, see Susanna Wade Martins and Tom Williamson (eds), *The Farming Journal of Randall Burroughes (1794–1799)* (Norfolk Record Society 58, 1995). In the next century some Lords Lieutenant, including Earl Brownlow in Lincolnshire, set their faces against farmers as J.P.s, not entirely on snobbish grounds. (See also R.J. Olney, *Rural Society and County Government in Nineteenth-Century Lincolnshire* (Lincoln, 1979), 98–103.)

As a magistrate Thomas was elevated from the rank of 'gentleman' to that of 'esquire', and acquired a certain status in the county as a whole. He attended county meetings from time to time, and in 1793, along with Tomline, sat on the Grand Jury for the March Assizes.[53] Nearer to home he took a prominent part in the Caistor Association for the Prosecution of Felons, whose membership included a number of those farmers and Caistor tradespeople who called on his judicial services from time to time. During the 1790s the inhabitants of the Caistor district were concerned not just with sheep-stealing and missing servants but with wider questions of public order and national defence: at this period England was threatened both with domestic unrest and with the risk of invasion from abroad. Dixon was on the committee of the Constitutional Association formed at Caistor in January 1793, and, as a subscriber of £20, on the committee of the county-wide movement for the formation of yeomanry troops in 1794. In November 1796 a mob at Caistor, consisting, it was reported, mainly of farmers' servants, disrupted the efforts of the Deputy-Lieutenants to enrol men in the supplementary militia, tearing up the lists of names and crying 'God save the King, but no militia!' At such times an active local magistrate could be very useful.[54]

* * *

Thomas seems to have had no idea of retirement, and was still in harness when he died in September 1798. A brief notice in the county paper, the *Stamford Mercury*, claimed that 'in him the poor have lost a real and good friend, and the neighbourhood an useful member'.[55] Those who came up before him as a magistrate may not have found him very friendly, but that he was a valued member of his local community there can be little doubt.

---

[53] *Stamford Mercury*, 15 March 1793.
[54] *Ibid.*, 18 January 1793, 13 June 1794, 18 November 1796.
[55] *Ibid.*, 14 September 1798.

4

THE OLD-STYLE FARMER: WILLIAM DIXON (1756–1824)

William Dixon was born at Great Limber, grew up at West Firsby and Riby, and went to school at Laceby and Brigg. But from his early twenties until his death he lived at Holton-le-Moor, and from 1785 he occupied the house there in which his descendants live today. This was the house which, as described in Chapter 3, his father provided for him on his marriage. The builder was Thomas Warmer of Caistor, who had previously done some internal work on the Riby house.[1] He began work at Holton in 1783, but the house was not finished until two years later, delayed possibly by his bankruptcy, and according to family tradition the William Dixons were still in the old house when their eldest son, Thomas John, was born in June 1785.[2]

Holton Hall, as it would be known, was erected on a site a little to the south-east of the old manor house, and the attractive bricks used in its construction were made from clay most probably dug only a few yards away. It was a handsome but unpretentious building of five bays and three storeys, unadorned except for a string-course and a textbook door-case. There was a dining room to the right of the small entrance hall and a parlour to the left, with an office behind the stairs and a long back wing for the kitchen quarters.[3] It was the kind of house, as William later remarked, whose occupant would be expected to keep a chaise and a liveried servant, but it was still a middle-class house rather than the mansion of a squire. It closely resembled Holly House in Caistor, also built by Warmer, and, less closely, Boundary Farm at Limber, built by the Brocklesby estate in the 1770s and later occupied by Richard Walkden.[4] Trees were planted round

---

[1] DIXON 22/8/1, fo. 80.
[2] For Warmer's bankruptcy, see notes on deeds for the Tower House, Caistor (Lincs. Archives, MCD 707).
[3] Plan and elevation in private possession. For sketches, see DIXON 18/6/1/1–2; 3 DIXON 5/1..
[4] Lincs. Archives, YARB 5/2/17/1. The house is dated 1773, but the estate accounts suggest that it was not ready for occupation until 1775.

the new house,[5] and a gravel sweep led to the front door. But otherwise its surroundings were less elegant. Immediately behind it were a dove house and various farm buildings, and in front it looked out over an expanse of bracken and furze. Miss Hickman, of Thonock, near Gainsborough, in a comment that reached the ears of the Ravendale Parkinsons, declared that the house 'might just as well have been built on Lee [Lea] Moor, some wretched place near Gainsborough'.[6]

The immediate drawback for William was not the view but the fact that he lacked a middle-class income to support this kind of establishment. Hall Farm, the holding that went with the house and in which his father had started him off, amounted only to about 107 acres excluding the warren.[7] His wife brought £1,000 when they married in 1782, and his father settled £100 a year in rental income on them, but the house had to be furnished, and he had to pay his father £25 a year for it. It was no wonder that he had to borrow £250 in the first year of his married life.[8]

He needed, therefore, to assemble enough land to make a viable holding, a project in which, it must be said, he received only minimal help from his father. In 1784 he was able to take over Barkworth's farm and part of Hall's farm (late Jacklin's); and around the same time he began to rent the Keelby farm and three closes of the Claxby glebe, bringing his total holding (again excluding the warren) to nearly 300 acres.[9] In 1789 he secured the tenancy of Ewefield, the farm that his grandfather had occupied, giving him at last some good-quality pasture at no great distance from the Hall, but he was obliged to borrow £500 at the time of the entry.[10]

All this was rented land, and apart from Ewefield was let to him by members of his family. But in the later 1780s he also began to *buy* land on his own account. This was a reversal of his father's policy, and harked back to the acquisitions made by his grandfather; but whereas old William Dixon had bought land as a way of investing money already realised by his farming and grazing business, young William made his purchases at least partly to further the family enterprise at Holton, and to do so he was forced to borrow money in anticipation of future profits. In 1789 he purchased from Cornelius Stovin 95 acres in Holton in the occupation of Michael

---

[5] The oldest trees were said in 1984 to be about two hundred years old (information from Mr Philip Gibbons).
[6] DIXON 16/4/10, Parkinson correspondence.
[7] DIXON 4/1, fo. 91; and see above, Chapter 3.
[8] 3 DIXON 5/11/7.
[9] DIXON 22/8/2.
[10] DIXON 1/A/2/25; 4/3, fo. 10.

Wiles, including 64 acres of the Breamer, a largely unimproved area of moor in the north-west of the parish.[11] (See Map 2.) He paid only £765 for the property, an indication of the poor quality of much of the land, but how he raised this sum is not recorded.[12] This transaction was followed in 1792 by a much more substantial one, when Stovin sold to William the remainder of his Holton property, composed of Wiles's house and buildings in the village, a cottage and some adjacent closes west of the village street, the Old Ground in the south of the parish, and the excellent farm known as Mount Pleasant. This holding, lying north of Ewefield, accounted for about 250 of the 319 acres that comprised the entire purchase, and represented a good proportion of the price of £6,400, the largest sum yet laid out by a member of the family in a single conveyance. William borrowed £500 from his father and £1,000 from his father-in-law, but left the remainder with Stovin on mortgage.[13]

William claimed later that these purchases were made 'for the sole accommodation and benefit of the premises I occupied of my father',[14] but they also enabled him to bring his farming business into a much more satisfactory form. He was able to occupy most of this newly acquired land himself: the fact that Mount Pleasant came with vacant possession was a particular advantage. (The outgoing tenant was Edward Young of Normanby.) He could now discard his outliers at Keelby and Claxby and concentrate his efforts at Holton, where all the land he held was a short walk from his own front door.[15]

William was also able to extend the family property into the neighbouring parish of Nettleton. In 1788 he bought a cottage there from John Turner, the Caistor attorney, and the following year he acquired a cottage and some 38 acres of open field land from the Barkworth family.[16] These purchases paid handsome dividends when Nettleton was enclosed in 1792–4. He obtained an allotment of 132 acres that was neatly arranged in a narrow strip along the eastern and northern sides of Holton parish, adjoining Dixon land and enhancing the security of the warren. The enclosure

---

[11] Stovin had Caistor connections, and sat on the Caistor Bench, but at this period was concentrating his resources on his estate at Crowle.
[12] DIXON 1/A/3/6. William's accounts are very incomplete before 1791, when he started his big account book (DIXON 4/3).
[13] DIXON 22/8/2; 4/3, fo. 203. See also Map 2. The house at Mount Pleasant bears the initials TW and CS (for Thomas Willoughby and Cornelius Stovin) and the date 1777.
[14] 3 DIXON 5/11/7.
[15] He also rented a small quantity of land in Claxby, on the Holton border, from the Brocklesby estate (DIXON 4/3, fo. 104).
[16] DIXON 1/C/1–2.

process cost him £300, followed by the expense of erecting buildings at Stope Hill, beyond the warren; but he sold another small allotment for £100, and recouped some more of his outlay by contracting for the boundary fence between the two parishes. He also added another 80 acres to his farming portfolio, using Stope Hill as a supplement to Hall Farm.[17]

* * *

The end result of this programme of expansion was to create a farm in Holton and Nettleton that was equal in size to his father's. Strictly speaking it was three farms, Hall Farm, Mount Pleasant and Ewefield, each with its own set of buildings, but William's flock was under a single shepherd, and his labour force could be moved around according to the priorities of farming work. The flock – 589 sheep in June 1795 – was a major source of profit, but Holton also grew more corn than Riby, and during the 1790s William increased his arable acreage, paring, burning and ploughing up some of his poorer pasture at Stope Hill and on the Breamer. In 1795–7 the buildings at Hall Farm and Mount Pleasant were improved. He also made the warren more productive, defending his rabbits from the poachers and taking advantage of the high prices of the period: in 1797 the skins and carcases of 422 dozen coneys fetched £300.[18]

Such an undertaking involved William in numerous dealings with his employees and neighbours. He had eight or so farm servants and labourers in regular employment, more than his father had at Riby, and his cottage and smallholding tenants were also drawn in from time to time. Beyond the boundaries of Holton his most frequent contacts were in the neighbouring parishes of South Kelsey, Nettleton, Claxby, Owersby and Thornton. More occasionally he transacted business with farmers at a somewhat greater distance, but still within the Caistor market area. These included George Nelson of Limber, Richard Goulton of Croxby and William Torr of Riby, all acquaintances of his father. As regards his largest accounts, however, and especially those for corn and rabbit skins, the merchants of Brigg featured more prominently than those of Caistor. This trend was accelerated by the building of the Caistor Canal in *c.*1794–7. William's farms were

---

[17] DIXON 4/3, fo. 203; Rex C. Russell, *The History of the Enclosures of Nettleton, Caistor and Caistor Moors, 1791–1814* (Nettleton, 1960). In contrast with Dixon's convenient allotment, the rector received some of the most hilly and awkward land in the parish. One of the commissioners was Samuel Turner of Collingham, a cousin of the Caistor Turners who through them would have been known to William Dixon.

[18] DIXON 4/3 *passim*. The prosecution of a poacher at Louth Quarter Sessions (6 April 1785) may indicate a more vigorous approach to the management of the warren.

within easy reach of the canal head at Moortown, from where his produce could be carried down to the Ancholme and thence to Brigg. He invested £300 in shares in the Canal, and his father £400.[19]

In 1796 sales of livestock, corn, wool and rabbit skins brought in a total of £1,000, but against this there were bills for labour, stock, seeds, buildings and other things, and fixed charges in the form of rent or mortgage interest, taxes, rates and tithes, bringing his net profit down to perhaps £150 or less. As well as a tenant and an owner-occupier, however, he was also a rentier landlord. Rents and interest on securities brought in about £285 a year gross, or £145 net after deduction of interest payable on his debts.[20] In a good year, therefore, he could reckon on a total net income of a little under £300. His investment in the farm, and particularly in his purchases of land, was paying off, helped by a period of high prices. But he still had heavy liabilities, and these would increase as his family grew.

The birth of Thomas John in 1785 was followed by those of Frances Martha in 1786, Amelia Margaretta (who died aged six) in 1787, Marmaduke in 1788, James Green in 1789, Rachel Harriet in 1791, William (who died an infant) in 1793, and lastly another Amelia Margaretta in 1794. Of the younger boys, Marmaduke may have been named after Marmaduke Tomline, whilst James Green's names alluded to his Parkinson grandmother's having been born a Green. William had to consider what would happen if he died while his affairs were still in a vulnerable state. In 1791 he made a will, leaving his wife £50 and the contents of the Hall – she was otherwise 'sufficiently and handsomely provided for' – but directing his executors and trustees (his brothers-in-law William Etherington and Robert Parkinson) to sell his freehold property in Holton and Nettleton in order to provide for his children's education.[21]

* * *

It was only a few years later, in December 1797, that William's father Thomas made his own will. We do not know who drew it up for him – perhaps John Turner of Caistor – but it was a more sophisticated document than had hitherto been usual in the family. He regarded the claims of his sons as already satisfied, but he was concerned that his wife, who had had

---

[19] DIXON 4/3, *passim*. See also Christopher Padley, 'Caistor Canal', *Lincolnshire History and Archaeology* 44 (2009), 5–22. This short length of canal prefigured some of the rural branch lines of railway in the next century in that it was under-capitalised and under-used. The shares were never of much value to the Dixon family.

[20] *Ibid.*

[21] BPD 6/11.

no formal settlement on her marriage, should be well provided for, and he wished his four girls to have handsome portions. Since he was not going to leave a large amount of ready money, these claims would have to be met by a distribution of some of his landed property. He therefore left the two farms in Thornton-le-Moor to his two elder married daughters, Gravel Hill to Rachel (Mrs Etherington) and Beasthorpe to Jane (Mrs Kelk). The Normanby estate was to go to Mary, and the sum of £3,000 (the equivalent in money of the other bequests) to Ann. Of this sum, £1,500 was to be charged on Beasthorpe and £1,000 on Gravel Hill, indicating probably that those sums had already been paid at the time of his daughters' marriages. Only the remaining £500 would come from Thomas's personalty. As for his wife, she was to have for her life all the property in Holton that he had power to leave by will – that is, Hall Farm and the warren – with remainder to William.[22]

These arrangements then had to be modified in two respects. Mrs Dixon, who had intended that her Keelby farm should pass to her eldest daughter Rachel Etherington on her death, changed her mind and decided that it should instead go straight to Rachel after Thomas's death.[23] And Ann married Richard Roadley, Thomas giving her £700 at the time. He therefore made a codicil to his will in July 1798. In view of the arrangement with respect to the Keelby farm he revoked the bequest of Gravel Hill to the Etheringtons and left it to his wife instead. This had the effect of increasing her likely income in her widowhood, including his money and securities as residuary legatee, to around £350 a year. And he reduced the value of the bequest to Ann to take the £700 into account.

His mother and sisters at Riby had encouraged William to expect £10,000 under the will, so when his father died not long after making this codicil he received a nasty shock. He had struggled to make his way at Holton without ever receiving, he claimed in a memorandum written at the time, more than £50 a year from his father, whereas his two clerical brothers had received 'near £300 pr ann each in education, money and preferment'. 'And now the fatal hour arrives, by which I find myself involved in despair my father's dead and left me nothing, therefore believe I am almost overcome with grief and disappointment.'[24] It was not strictly true that he had been left nothing, but he would not come into the absolute possession of

[22] DIXON 2/1/2.
[23] 3 DIXON 5/11/8. She had previously made a settlement of the Keelby farm, with Tomline as a trustee, but it had left her with a power of appointment.
[24] 3 DIXON 5/11/7.

Hall Farm or his house until his mother's death – and she was to live into her eighties. The blow was made heavier by the manner of its delivery. It was plain that he had not been in his father's confidence: Thomas had named his widow as his executor rather than his eldest son, and the will expressly directed that William was to continue to pay to his mother the interest on the £1,000 that he had been lent by his father.[25] What William had been hoping for above all was some ready cash. 'Money appears to be more preferable than land', he wrote at the end of his private outburst, and for a time he even contemplated selling some of his land before realising that he could not do so with 'propriety'.[26]

Although he took no such drastic measures the experience had a profound effect on him. From then on he turned away from personal ambition and concentrated his energies on less worldly objects. The next chapter will discuss his religious and philanthropic thought, and how his philosophy affected his public life. The remainder of this chapter will trace his gradual retirement from business, and his transformation from an enterprising young farmer into an increasingly old-fashioned one.

First, however, one further consequence of Thomas's death needs to be considered. It had been agreed – perhaps at the time of Ann's marriage, perhaps before it – that Richard Roadley should take over the tenancy of the Riby farm. It would thus remain in the family, and Mrs Dixon would be able to stay in the house where she had lived for the previous forty years. But in 1803 Marmaduke Tomline died. His heir, Bishop Pretyman, called in the well-known north Lincolnshire agent and valuer John Burcham, who recommended a thoroughgoing reorganisation of the Riby holdings. Roadley's farm was reduced in size, and this may have prompted him to consider a move.[27] In 1805 an opportunity presented itself when the Searby estate, near Brigg, came on to the market.[28] It had a good mixture of wold and low-lying land, and Roadley planned to farm most of it himself. The only drawback was the lack of a suitable house, but a new one was put in hand. The family, including old Mrs Dixon, was transferred temporarily to Laceby while it was being built, but moved over to Searby

---

[25] In fact William's mother let him off these payments, thus reducing her own income by £50 a year.
[26] 3 DIXON 5/11/7.
[27] LIND DEP 29/5; DIXON 3/7/15.
[28] For the sale catalogue, see DIXON 20/1/1. The estate was mostly freehold but included some land leased from the Dean and Chapter of Lincoln. Roadley left part of the purchase price on mortgage with the vendor, Edward Weston of Somerby (DIXON 3/7/215).

in 1807.[29] Roadley's new domain lay almost as near to Caistor as it did to Brigg, and – a fact that was to become of great significance in the next generation of the Dixon family – it was only five miles north of Holton.

* * *

Meanwhile William Dixon had his sons' education to consider. Christopher Cave, who had taught him at Brigg, had moved to Caistor Grammar School, and this may well have influenced him in sending his sons there. Thomas John attended from 1798 to 1801, Marmaduke from 1803 to 1805, and James Green, his youngest and slowest son, from 1803 to 1808.[30] Thomas John, or Tom, showed an early aptitude for business, perhaps taking after his great-grandfather, and there seems to have been no doubt that he would become a farmer. James Green, although a very different character, also showed a bent for farming. Given his clerical relatives it might have been expected that Marmaduke would enter the Church, but William had before him the discouraging example of his two brothers, and in any case Marmaduke himself showed no inclination for the life of a parson. It was decided instead that he should be trained as a solicitor.

In launching Tom and James on their farming careers William treated them as apprentices and then as junior partners before establishing them in their own farms. In this he was following his grandfather's practice rather than his father's – but with this difference, that he had two sons to consider rather than one. By 1800, although still at school, Tom was already helping his father with tasks such as dragging turnips, mending rabbit nets and riding round the warren. He was also introduced to the mysteries of accounting, as far as William understood them himself. From 1801, now aged sixteen and back from school, Tom began to take on more responsibility as a kind of bailiff, dealing with the workpeople, keeping tallies of corn and livestock and sometimes attending the Caistor and Lincoln fairs.[31]

In 1801 William took advantage of an opportunity in Nettleton. Richard Roadley, who had acquired the advowson of the living, presented his brother-in-law William Jackson to it, and in return Jackson leased his glebe

---

[29] DIXON 9/6/5/12.

[30] The two younger boys were not under Cave but under the Revd Rowland Bowstead, who succeeded Cave as Master in 1802 (David Saunders, *The Story of Caistor Grammar School Lincolnshire from 1631 to 1945* (printed Heighington, 2007)).

[31] For Thomas John's early accounts and day books, see DIXON 5/4/1; 5/6/1–2; 9/1/1–5. For William's own difficulties with accounts, see DIXON 4/3, fo. 30, where he observes that in a 'tripartite business', that is where three parties were involved instead of the usual two, one could get into 'great perplexities'.

to Roadley. It consisted of nearly 400 acres, much of it allotted in lieu of tithes at the enclosure, and William took 185 acres as Roadley's under-tenant. The fields lay not far to the north of Stope Hill, and could be farmed with the other Nettleton land on the other side of Holton moor.[32] On paper this was a further expansion of William's business, but the purpose was almost certainly to give Tom's energies more scope and to prepare the way for the next stage in his career. This took place in 1805, when William set him up in partnership with his uncle Roadley, in a holding that covered not only 250 acres of the Nettleton glebe but other Dixon land in Nettleton and Beasthorpe Farm in Thornton, a total of some 600 acres. As part of the arrangement William bought Beasthorpe from his sister Mrs Kelk; but he left the whole purchase price of £3,250 on mortgage with her, using the rent from the partnership to pay her interest.[33] (Thomas John took over the mortgage payments in 1812, and in 1818 took over the lease itself.) In 1807 Roadley withdrew from the partnership in order to concentrate on Searby, and it was adjusted to give William a one-third and Tom a two-thirds' share. The latter, however, had the sole management of the holding. He added to it some carr land belonging to Gravel Hill, and in 1810 took the rest of the Nettleton glebe, bringing his two-part holding up to about 700 acres, at a total rent of about £630 a year. In 1812 William withdrew from the partnership altogether, leaving Thomas John as the sole tenant.[34]

In 1806 or 1807 William wrote to a Yorkshire wool merchant, 'I really have no money to spare, my children's education and fixing in business take all my spare cash.'[35] In 1805 Marmaduke had entered the Market Rasen office of Messrs. Tennyson and Main, the leading legal firm of the district. He was articled in 1806,[36] and in 1810 went to London to gain further experience – an unhappy period when he pined for the green fields of Lincolnshire.[37] George Tennyson was willing to take him back into the firm, at least temporarily, but he was keen to be his own master. Fortunately an opportunity came up at Caistor. George Bulmer left the district,[38] and in January 1811 Marmaduke took over his practice, along with Bulmer's house,[39] his clerk William Moody and a portfolio of local clerkships. (One

[32] DIXON 4/3, fo. 192; LQS, Nettleton land tax duplicates.
[33] DIXON 4/3, fos 31, 110.
[34] DIXON 4/3, fo. 126.
[35] DIXON 7/1/5: 13 October 1807.
[36] The premium was a hundred guineas (DIXON 4/3, fos 46v, 55v).
[37] DIXON 9/2/11 *passim;* 9/6/1.
[38] Bulmer left under a cloud (DIXON 4/3, fo. 76v; *Stamford Mercury*, 24 August 1810).
[39] This was Tower House, built by Warmer.

of these, the Yarborough coronership, was in the gift of the freeholders of the wapentake. William issued an address asking for support for his son's candidature, citing his 'uniform exertions to serve the country [i.e. the district] in various capacities for near THIRTY YEARS PAST', but in the event there was no contest.[40]) William calculated that the total cost of educating Marmaduke and getting him settled in his profession had been £1,421.[41]

In 1808 James Green left school, his spelling still somewhat shaky, and began helping his father on the farms.[42] He was very different in temperament from Thomas John, and while the two lived at home (neither was to marry for some years) there was friction between them, with William inclined to take the part of James. 'It is no use,' he wrote in his journal, 'me as a Father using every effort to provide impartially and justly for all my children: if the elder take from the younger what he ought to enjoy my efforts prove all in vain.'[43] At around this time he contemplated putting the day-to-day management of Hall Farm and Mount Pleasant into the hands of his two farming sons jointly, but must have realised that that would not work, and that James would need a farm of his own.[44] A proposal was put forward, and supported by Mrs Dixon senior, to set him up in the large South Farm in Thornton, a holding of some 440 acres, but this appears to have been negatived by James himself.[45] In 1812, however, he took the tenancy of Mrs Dixon's Gravel Hill farm. A little under 300 acres, it was conveniently situated between Beasthorpe to the west and the Holton and Nettleton land to the east. The entry cost £1,224 11s, of which Thomas John contributed £857 and Mrs Dixon £140.[46] The tenants were James and his father, but James did the day-to-day work.

In 1814 James added another farm to his holding. William gave up Mount Pleasant, in Holton but adjoining Gravel Hill, and James entered it at a valuation of £962 19s 6d for the 216-acre farm. This time William lent two-thirds of the money, partly in the form of stock as the outgoing occupier, and Mrs Dixon senior the remainder.[47] In 1815, following her

[40] DIXON 7/1/8: 4 September 1810.
[41] DIXON 4/3, fo. 86.
[42] For James's early day books, see DIXON 12/5/33–4.
[43] DIXON 7/1/9: 27 October 1811.
[44] DIXON 7/1/9: 12 November 1811. Under the proposed arrangement William would have retained control over cropping and the management of the warren.
[45] DIXON 9/6/1: T.J. Dixon to ?Robert Parkinson, *c.* July 1811; DIXON 7/1/8: Francis Raynes (for whom see below, note 52) to William Dixon, 3 July 1811.
[46] DIXON 4/7, fo. 92v; 12/1/1, fo. 96; 12/1/2.
[47] DIXON 12/1/1, fos 98v, 99.

had made a very appropriate match. The Skipworths were not only near neighbours, but had a similar background to the Dixons: her new father-in-law Philip (1745–1825) had come from Aylesby, where he had been a successful farmer and sheep-breeder, and at one time his younger brother Thomas had held one of the Riby farms.[60] But in the last twenty-five years he had risen into the ranks of the landowners of north Lincolnshire in a much more spectacular fashion than had the Dixons. He had purchased 73 acres in Rothwell in 1793, 250 acres in North Kelsey in 1795 and 60 acres in Caistor, with two public houses, the lordship of the manor and the right of free warren, in 1802.[61] But in 1802 he also entered into a partnership to buy the lordship of South Kelsey. It had been enclosed in 1797, but was capable of considerable improvement, and extended to as much as 4,500 acres. The original intention of the partners in this speculation, who included George Tennyson at one stage, was to sell on the estate at a profit;[62] but Philip Skipworth increased his stake in 1804, and in 1808 became the sole proprietor, with the aim of farming much of it as a family business.[63]

Despite this great accession of property Philip remained a plain north Lincolnshire farmer. (In March 1818, writing to his friend and business associate George Tennyson about his attempts to arrange a meeting with a third party, he explained that the man in question 'left home Saturday last and don't return whilst Thursday'.[64]) But he made a gentleman out of his third son George, sending him to Oxford and eventually making him his principal heir. Philip occupied a modest house in South Kelsey when he settled in the parish (the big Tudor hall having been pulled down), but built a new house for George at Moortown, only a couple of miles north of Holton. The situation was not perhaps of the most picturesque, but the house had some pretensions: it boasted *two* bay-windowed drawing-rooms.

---

[60] Philip's father had moved to Aylesby from Alvingham, a parish on the Middle Marsh near Louth where the family had been settled since the seventeenth century. See also Appendix 2, Table 4.

[61] DIXON 1/G/1; 1/H/1. Amelia's jointure was secured on the Rothwell estate in 1815, but transferred to the Caistor and North Kelsey properties in 1825 (DIXON 2/4/1; 1/H/4).

[62] For an earlier example of a speculative purchase by a consortium of farmers, see the Humberston estates (for which Thomas Dixon had been steward), sold on to Robert Smith (first Baron Carrington) in 1792 (Centre for Buckinghamshire Studies, Carrington Papers, Box 33). For the involvement of George Tennyson in various land deals in the late eighteenth and early nineteenth centuries, see Lincs. Archives, TDE F/1–2; 4 TDE F/6.

[63] DIXON 1/F/1/9–15.

[64] 4 TDE F/6.

By 1825, when his father died, George was farming nearly 1,000 acres, and had been confirmed as the heir to his father's entailed property.[65]

With all his children now settled, William could subside into semi-retirement, during which he continued to occupy Hall Farm and the warren. In fact he surrendered them to Thomas John only in 1823, a few months before his death. But in these latter years his affairs were not in an entirely satisfactory state. Around 1820 he had a gross income of around £900 a year, comprised principally of rents from his farming tenants, but out of this sum he had to find £576 a year to service the interest on his debts. 'I owe a great deal of money, I do not hesitate to acknowledge,' he wrote, 'but I hope I have very great plenty to pay all and everyone, and something handsome to spare and appropriate to charitable purposes.'[66] This was unduly optimistic: over the years he had, as we shall see, already diverted quite a bit of his wealth to charitable purposes, and his income was falling off, the effect of the post-war depression and a slump in the rabbit-skin market.[67]

Caistor Moor was enclosed in 1811–14, during the period of high corn prices, but Holton Moor remained its old scrubby self, with its rabbits munching the nearby crops and even getting into the church to nest among the surplices.[68] William's devotion to his warren amused his neighbours and exasperated his son Thomas John, who saw that it could be put to more profitable uses even in the post-war years.[69] There had in fact been no less than four attempts to enclose the moor, in 1807, 1813, 1814 and 1817, but they had all failed.[70] It is not known who the prime movers were on each occasion, but the attempts of 1813 and 1814 are likely to have been related to the tithe question. Between 1810 and 1814 William was in dispute with Lord William Beauclerk, of Redbourne Hall, the lessee of the prebend of Caistor, who at one point was claiming the vicarial as well as the rectorial tithes of Holton. An allotment of land in lieu of tithes at the enclosure of the moor would have removed a source of friction and

---

[65] DIXON 2/4/2. There is a parallel, though not an exact one, in the Tennyson family, where George, although not 'disinheriting' his eldest son, chose his third son Charles (later the Rt Hon. Charles Tennyson d'Eyncourt) as heir to the Tealby estate.
[66] DIXON 7/3/9, fo. 27.
[67] DIXON 3/1; 4/3, fos 41, 68–9, 84, 263.
[68] DIXON 7/1/9: 23 July 1811.
[69] DIXON 7/1/14: 23 and 25 November 1818.
[70] Richard Olney, 'The Enclosure of Holton-le-Moor', in Dinah Tyszka, Keith Miller and Geoffrey F. Bryant (eds), *Land, People and Landscapes: Essays on the History of the Lincolnshire Region written in honour of Rex C. Russell* (Lincoln, 1991), 121–4.

contention in the parish.[71] But legal complications as well as conflicting interests may have stood in the way of an agreement. William himself had mixed views about the benefits or otherwise of enclosure generally;[72] but at Holton it would have meant the end of his warren, and that for him had perhaps been the sticking point.

* * *

The forty years during which William Dixon farmed at Holton had their economic ups and downs, but as a whole they were a period of prosperity for the agriculture of north Lincolnshire such as had not been seen since the great days of sheep farming in the early eighteenth century. A few exceptionally successful farmers, such as Philip Skipworth and Richard Roadley, were able to purchase whole estates that had previously been owned by country gentlemen.[73] Other farmers did not go in for land purchases on that scale, but during the same period their wealth and standard of living greatly increased. Around 1823 George Nelson of Limber told the Revd John Parkinson that 'his predecessor gave only £100 a year for his farm at Limber and died insolvent. Mr Byron's father gave the same sum and … was very hard set. Yet they had but two pot days in the week at that time and seldom any bread in their houses but what was made of barley and oatmeal.'[74] A generation later the occupants of many farms of a similar size were in very comfortable circumstances.

William Dixon benefited from these changes but did not wholeheartedly embrace them. In his farming he was slow to move with the times, at least after 1800, and his purchases of land were incremental rather than spectacular. But the *family* enterprise over which he presided at Holton was impressive in its way. Around 1813 he was farming 500 acres himself, but he had started James Green on 300 acres and Thomas John was already occupying 700. The area in family occupation stretched over six miles, from the Ancholme to the Wolds. Ten years later the family holding had increased to 2,100 acres, made up of 1,400 in Thomas John's hands and 700 in James Green's. In this sometimes complex process partnerships, formal or otherwise, had been important – Thomas John's with his uncle Roadley and James Green's with his father. So had the money and influence of Mrs Dixon senior. In her lifetime she helped to get James started,

---

[71] DIXON 4/3, fo. 12; RED 2/4/6.
[72] See below, Chapter 5.
[73] South Kelsey had been owned by Francis Foljambe of Osberton, Nottinghamshire, and Searby by Edward Weston of Somerby, near Brigg, the next parish to Searby.
[74] DIXON 16/3, p. 265.

and her death enabled William to continue his withdrawal from active farming in order to concentrate on his role of landlord.

As in his farming, so in his way of life William looked askance at new-fangled ways. Holton Hall had been built before the age of architectural refinement in north Lincolnshire, and he was not tempted to add bay windows in emulation of Searby Manor or Moortown House. His wife had no farm servant in the house, but she continued to make do with only two living-in maidservants.[75] She may even have had to wait until 1808 for a chaise.[76] She seems to have been more sociable than her husband, and she probably appreciated the comforts of her daughter's establishment at Moortown, but there is no evidence that she had social *ambitions*. Her husband certainly took no pride in Amelia's elevation, and in general he liked to present himself as a man of unostentatious and indeed homespun habits. He favoured plain woollen garments, for instance, rather than the more fashionable cotton ones.[77] A story was told of him (as it was of other rural eccentrics) that he sometimes wore a smock, and that in this garb he was once mistaken by a visitor for one of his labourers. (Unfortunately there is no portrait of him extant. He would no doubt have considered it a vainglorious waste of time to sit for one.) He saw himself as 'a keeper of sheep and tiller of the ground', and discouraged people from addressing him as 'esquire'.[78] He probably retained a pronounced north Lincolnshire accent: his son James Green, who took after him in many ways, certainly did. In his younger days he was not averse to the occasional drink, but after his father's death he became more puritanical.[79] In 1805, for instance, on a visit to Mr Freer of North Kelsey, he was prevailed upon to consume pigeon pie during Lent: 'if I had been at home my wife would have reminded me by producing proper victuals'.[80]

As he grew older he became more averse to noisy social occasions, as well as to the expense that they entailed. In 1812 he lamented 'the evil

---

[75] DIXON 4/3, fo. 266v.
[76] DIXON 4/3, fo. 23; 7/1/9: 14 April 1811.
[77] DIXON 7/1/5: 29 July 1807.
[78] DIXON 7/2/9; 7/1/11: 29 September 1813 – a reused letter of 1787 in which the writer apologises for addressing him incorrectly, 'but understood you were usually addressed in the same manner'.
[79] DIXON 7/1/11: September 1813 – a reused letter from T.H. Swan, issuing an invitation and adding 'I know you are a Tipler'.
[80] DIXON 7/1/1: 27 March 1805. William's wife remains a shadowy figure, but she seems to have been strongly religious. In 1810 Mrs Parkinson of Ravendale wrote to her commiserating on the death of her daughter Frances, but saying her 'great piety' would doubtless sustain her (DIXON 13/2/5/4).

customs now of training up young people to splendid and long meals, their desserts with wine and fruits waste a great part of the afternoon or evening'.[81] In 1809 he had been uncomfortable when his sister Roadley paid a rather formal visit to Holton accompanied by Mr Roadley senior, Mr Etherington and Mr Hudson, the vicar of Searby.[82] Some years later he became restless when his wife and son were entertaining 'two acquaintances who were diverting themselves with music', and wished afterwards that he had left the party to go and mend rabbit nets or visit his old warrener.[83] As for going out in order to enjoy himself, there is one (probably untypical) reference to his joining the officers at Caistor to celebrate the King's birthday in 1806,[84] but he regarded the Barton races and even Caistor assemblies as licentious affairs.[85] His pocket or appointment diaries, as opposed to his journals, are very incomplete, but there is no reference to his ever taking his family to the Lincolnshire coast or to a Yorkshire watering place, let alone going up to London. He was no sportsman, and his papers contain no reference to hunting, or to spending much time in the saddle.

Altogether he found it difficult to rub along with his fellow human beings, were they his labourers, his neighbours or, at times, members of his own family. Perhaps to compensate for this, he was good at communicating with himself. He had a great reverence for the written or printed word, and he liked to record his observations and thoughts. It is to the products of his reading and reflection that we turn in the next chapter.

[81] DIXON 7/1/10: 15 November 1812.
[82] DIXON 7/1/7: 27 March 1809.
[83] DIXON 7/1/14: 20 November 1818.
[84] DIXON 7/1/3: 4 June 1806.
[85] DIXON 7/1/8: 22 March 1811; 7/1/11: 28 November 1812.

5

WILLIAM DIXON AS PHILOSOPHER AND PHILANTHROPIST

William Dixon was a thoroughly religious man. As a child he had been taught to 'make God my friend', and had imbibed the doctrines and liturgy of the Church of England.[1] In his early years at Holton he had little leisure for matters outside the farm, but from around 1793 he turned his attention to the religious state of the parish. In that year he secured an increase in the provision for church services from £6 13s 4d a year to £13 6s 8d, enabling them to be held once a fortnight instead of only once a month. Also in 1793 he engaged a schoolmistress, sharing the expense with Robert Bett, the only other substantial farmer resident in Holton. To judge from her salary of seven guineas and a list of holidays, she was to preside over a day school, an ambitious venture for that date. It seems, however, to have lasted only one year, but thereafter it became a Sunday school, meeting in William's house until he provided a little school-room for it in 1822.[2] He encouraged psalm-singing in church, appointing a choirmaster in 1796, and set an example to the parish by adhering to a regular routine of family prayers.[3]

As described in Chapter 4, the circumstances in which William succeeded his father as head of the family in 1798 had a profound effect on him. He did not undergo a classic religious conversion, but experienced what he himself described as a 'regeneration'. He began to think much more deeply about spiritual matters, and set himself a programme of serious reading. But this did not lead to a reclusive obsession with the state of his soul. On the contrary, his reading and thinking extended from religious literature to political economy; and his good works, which before had been confined largely to his own parish, took on a broader scope. 'Since I have become regenerate', he wrote in 1805,

> I am always the happiest when I am going about doing good, and I take the greatest pleasure in spending my time to profit others as well as myself. I

[1] DIXON 7/3/2; 7/1/9: 27 October 1811.
[2] DIXON 4/3, fos 179–80. In 1824 it had 23 pupils (DIXON 7/1/21).
[3] DIXON 4/3, fo. 21; 7/3/1, 2, 6.

consider my service here is, or ought to be, to assist mankind, those who are in want or need, and when I am otherwise employed I am unpleasantly situated and not satisfied.[4]

His efforts to serve mankind during the last quarter-century of his life will be described later in this chapter, but first we take a closer look at his religious and social thought as it evolved after 1800. We can do this because he left a series of notes and other writings, in themselves an unusual survival for an early nineteenth-century farmer. Of most interest are his journals, which run from 1805 until his death in 1824, although they are less full for his later years. They are a product of his regeneration, and it is possible that he started them in 1798 or 1799, the earliest volumes having been lost; but it is more likely that only in 1805 did he find himself with sufficient leisure for their compilation to become a settled habit.[5] An early practitioner of recycling, he made up each volume from any scraps of paper that came to hand – letters, fragments of accounts or printed matter such as playbills, announcements of ram sales or circulars offering rewards for the apprehension of felons.[6]

The primary purpose of these notes was as a record of his private devotions. On weekdays he would rise early and devote a couple of hours before breakfast to prayer, bible reading and reflection. On Sundays he would transcribe the collect, and sometimes the epistle, and his meditations would form almost the materials for a sermon. Having been brought up to keep regular accounts, he saw this kind of record-keeping as a kind of spiritual audit (a not uncommon middle-class notion of the time).[7] 'How do we give up our account to God?' he wrote. 'Have we ever tried to settle annual accounts with Him? Can we show a specimen of our book-keeping satisfactory either to ourselves, our brethren of Mankind or any ways suitable to God's purposes for which we were created ...?'[8]

Besides the Bible, the Book of Common Prayer and the Homilies of the Church of England he explored a range of spiritual, devotional and theological literature, partly courtesy of the S.P.C.K. Some of it derived from as far back as the seventeenth century, and of the more recent writers

---

[4] DIXON 7/1/1: 27 March 1805.

[5] For the tendency of middle-class people to become more religious in retirement or semi-retirement see, e.g., Davidoff and Hall, *Family Fortunes*, 76, 145–6, 226.

[6] DIXON 7/1 *passim*. In this way the journals double as a valuable collection of contemporary ephemera.

[7] Margaret R. Hunt, *The Middling Sort: Commerce, Gender and the Family in England 1680–1780* (Berkeley and London, 1996), 174.

[8] DIXON 7/3/9, fo. 32 (a stray leaf from a journal?).

some had evangelical and even nonconformist connections. One such was David Simpson, an earnest 'Methodistical' preacher and the son of a Yorkshire farmer, and another was Job Orton, a pupil of the great Northampton minister Philip Doddridge.[9] But for William Dixon the civil constitution of the country was also part of the divine dispensation. For legal authorities he turned most frequently to Blackstone and to Burn's *Justice* (perhaps in his father's old copy), whilst in matters of political economy he consulted Locke, Adam Smith (though not always agreeing with him) and, in his later years, Ricardo.[10] If these sources began to give him mental indigestion he could always derive food for thought from a paragraph in the *Stamford Mercury* or *Bell's Weekly Messenger*, while in certain moods the recollection of an unsatisfactory encounter with a neighbour or workman could set off a chain of reflections. In short, as with many autodidacts, anything read or remembered could become grist to his intellectual mill.[11]

Dixon was no visionary like Blake, or polemicist like Cobbett. His vision of a life well lived was a limited and local one, closely connected with his everyday work. 'Judge and determine', he apostrophised himself,

> how to arrange and manage your daily farming business by comparative conclusions from experiments with facts. Use your Reason, your Religion, and conform to Law. In all your actions a combination of the three above mentioned processes are necessary for your well-being as a Man of business and a useful Member of society.[12]

Yet he could not fail to be affected by the times through which he was living. His journals are fullest for the years 1805–14, when England was at war with France, and like most of his neighbours he saw the military struggle as one to defend the nation's religion as well as its liberties. But he was conscious that in many ways the times were out of joint. Fortunes

---

[9] David Simpson, *A Plea for Religion and the Sacred Writings* (Macclesfield, 1797); Job Orton, *Sacramental Meditations* (Shrewsbury, 1777). For Orton, see Alan Everitt, *Landscape and Community*, 230.

[10] William was the first member of the Dixon family to have a working library of volumes on religion, agriculture, history and other topics, mostly very cheaply bound. A few were inherited from his father or the Walkdens, others perhaps from the Parkinsons (although some of those came to Thomas John from his uncle Robert Parkinson). These books were left to Lincolnshire Archives Office by George Dixon in 1970.

[11] Dixon's journals may be compared with the diary of a wold farmer of a later generation, Cornelius Stovin of Binbrook (see Jean Stovin, ed., *Journals of a Methodist Farmer 1871–1875* (London, 1982)). Like Dixon, Stovin was a thoughtful man, perhaps more comfortable in his study than in the day-to-day world.

[12] DIXON 7/1/8: 10 May 1811.

were being made by speculators, while high prices were injuring the poor. Altogether it was becoming harder to 'reconcile the laws of God and Man'.

* * *

About one matter, however, he was quite clear. Liberty and fraternity might be all very well, but equality was 'contrary to the will of the Almighty'.[13] He subscribed wholeheartedly, in fact, to the hierarchical model of society described in Chapter 1.[14] For him land was the basis of the social order, just as he regarded the price of land as the ruling factor in the local economy,[15] and it was natural to him to think in terms of the social order as comprising those who owned the soil, those who farmed it and those who worked it. Among the landowners he applauded those who understood the obligations that went with their wealth. The Duke of Ancaster, Lord Lieutenant of Lincolnshire, held no property in Dixon's part of the county, but when he died Dixon nonetheless wrote him an appreciative private obituary. 'Consider how many persons, parishes and families are made comfortable and happy by living on the charitable and fruitful domain of such a nobleman.'[16] In contrast he cited Goxhill, near Barton, as an example of a large open parish lacking landed influence, and full of needy small freeholders who contributed little to society.[17]

As for the farmers, he habitually wrote of them as *tenant* farmers, even though he himself and not a few of his acquaintance were owners as well as occupiers of land. He recognised that he had a more substantial business than many, and indeed regarded occupants of fifty to a hundred acres as small farmers.[18] But, large or small, the farmer must not get above himself. He 'should act as a person in an inferior Station to a landlord, tho' his circumstances be in a state of affluence or poverty'.[19] He should eschew both excessive ambition and luxurious idleness. As we have seen in Chapter 4 Dixon himself projected the image of a plain farmer; and he also had things to say about farmers' wives and daughters. The wives should not

---

[13] DIXON 7/1/10: 10 November 1812.
[14] He quoted with strong disapproval the views of Richard Barwick, a 'fiery zealot' who believed that there were only two classes, rich and poor (DIXON 7/1/16: 5 April 1820).
[15] DIXON 7/1/8: 22 April 1811.
[16] DIXON 7/1/7: 10 February 1809.
[17] DIXON 7/1/1: 14 March 1805; 7/1/3: 15 April 1806.
[18] In 1816 he noted that some small tenants of £100 or even £50 were surviving the post-war depression, whereas others, 'who live now in an age of improvement as it is called', had taken farms of up to £1,000 a year but were now losing money (DIXON 7/1/13: 13 May 1816).
[19] DIXON 7/1/9: 4 November 1811.

retire to their nearest market town when their husbands died, but should stay in the country to impart their dairying skills to the next generation, while their daughters should not be too proud to help in the house and around the yard (just as shopkeepers' daughters might be expected to take their turn behind the counter).[20] Finally, as the farmer should rest content with his situation, so should the labourer. Farmers should set a good example to their men, and had only themselves to blame if they did not do so: 'If we farmers give over working we shall soon find our servants get proud and self-sufficient.'[21]

When he looked round his own neighbourhood he found as much to deplore as to praise. Among the magnates he had a closer view of Lord Yarborough than of the Duke of Ancaster, and had mixed feelings about the former: he was a conscientious landlord, accommodating to his tenants and pleasantly condescending to his neighbours, but he spent a good deal of his time out of the county, and sometimes failed to give a lead in local affairs.[22] For Lord William Beauclerk, the grasping lessee of the Caistor prebendal estate, he had no time at all: such arrogant conduct created 'murmurs and revenge instead of content and unanimity'.[23] Among his farming neighbours he particularly deplored the custom of purchasing large estates. 'We farmers by becoming purchasers of land injure the community exceedingly. We grow high and proud and are not content to be tenants but want to exalt ourselves ....'[24] The possession of a freehold might bring security, but it could not make its owner a gentleman if 'breeding and cultivation were lacking'.[25] There was of course one conspicuous example of this just the other side of Dixon's parish boundary. He kept by and large on neighbourly terms with Philip Skipworth, but there are indications in the journals that the purchase of South Kelsey had put his nose out of joint. He commented, for instance, on the strain on local credit that it had caused.[26]

---

[20] DIXON 7/1/2: 10 February 1806; 7/1/3: 8 July 1806.
[21] DIXON 7/1/2: 20 February 1806.
[22] DIXON 7/1/2: 2 January 1806; 7/1/8: 12 August 1811; 7/1/1: 23 March 1805; 7/2/12. Yarborough's help in preventing a general collapse of credit following the failure of Marris and Nicholson's Barton bank in 1813 was appreciated, but, as noted later in this chapter, his failure to encourage the Caistor Society of Industry was a mark against him.
[23] DIXON 7/1/1: 17 March 1805; 7/1/8: 9 August 1811.
[24] DIXON 7/1/1: 14 March 1805.
[25] DIXON 7/1/11: 11 July 1813.
[26] DIXON 7/6/5–8. The South Kelsey purchase prompted William to draft a pamphlet called 'Anti-Anthropopagi, or a Serious Address to the County on the conduct of a certain Clan ...of his Majesty's subjects denominated Land Jobbers', in which he hoped

William Dixon and his family had also been purchasers of land, although not on such a scale, and he was not altogether at ease with his own role as a landowner. He was conscious that he could be criticised for charging a high rent for Jacklin's, but then it had been bought when land was dear, and had now to support his wife's jointure.[27] Then there was Mount Pleasant, for which he himself had paid a large sum: he considered however that he owed it to his station in life, his family and 'the community where I live' to make good the investment.[28]

It is noticeable that when discussing his social equals he shows no sense of middle-class solidarity. He nowhere uses the term 'middling sort', let alone 'middle class'. The clergy he regarded as a caste set apart by their divine calling, although that did not prevent him from criticising parsons who were too fond of worldly pursuits or too eager to exact their full quota of tithes.[29] He was 'anxious to see the time when [the clergy] will be as much respected by the Laity as a good Landlord is at this present time by his Tenants'.[30] He was even more critical of the agents, valuers, land jobbers and the like who did well out of the war. He shared Cobbett's dislike of speculators and those who merely circulated paper money instead of doing productive work.

Where he was most ambivalent was when it came to the labouring poor. Towards his own men his attitude had a strong element of paternalism. He regarded them as members of his 'family', even though none of them lived under his roof; and he took the view that in general the living-in system for farm servants made it easier for a master to exert the necessary moral influence over his men.[31] 'Most labourers want as much minding and observing as children at school,' he wrote, and he worried over their tendency to get themselves into difficulties over small financial transactions. Although he showed little interest in the quality of their cottage accommodation he was concerned that they should have enough to live on, especially in the years of dear bread. The Holton labourers kept cows, and were allowed

---

that the farmer involved, who was reported to have a good opinion of the lordship in question, would give up his other farms and take the new purchase in hand rather than break it up for sale. He showed the draft to the Revd Samuel Turner, but took it no further.

[27] DIXON 7/1/5: 1 November 1807.
[28] DIXON 7/1/6: 30 January 1808.
[29] In his later years he conceded that some Methodist preachers were more zealous and eloquent than their Church of England counterparts, although he professed to detect more failings in their followers than in the Anglican laity (DIXON 7/3/9, fo. 71 [c.1820]).
[30] DIXON 7/1/6: 30 January 1808.
[31] DIXON 7/1/2: 2 February 1806; 7/1/11: 11 September 1813.

access to grazing for them, although during this period they appear to have been restricted to one cow each rather than the customary two. In his later years he became converted to the value of allotments. He held that enclosure deprived the poor of their interest in the soil, and in 1824 drafted a letter to the Ulceby enclosure commissioners advocating the provision of a cow pasture.[32] As we have seen, he also believed strongly in the merits of working-class education. Not only did he emphasise its religious and moral value, but he saw practical benefits in having a literate and numerate workforce.

> Why can it be disadvantageous to the community or why should the ministers of the Church of England oppose the educating of children on a Sunday in writing and accounting? It can in no ways impare the industry of the labourers or impede the labourer from doing his duty, but must add to his comfort and convenience and strengthen his faith and confidence. I have 3 or 4 labourers can write and read very decently and figure an account, and they are very trusty, mindful, industrious and hard working men, and superior on that account to any I ever had without or destitute of that education.[33]

In his dealings with his men, however, his enlightened views could come up against his everyday experience as an employer, and his relations with his men were not always easy. He notes the self-conceit of his shepherd, and on another occasion overhears his warrener 'disclaiming' against him. No amount of paternalism could conceal the divergent interests of capital and labour, and there were times when he felt his kindness unrewarded and his authority threatened.[34] Just as some farmers wanted to exalt themselves into landlords there were labourers who aspired to become farmers. In February 1809, thinking of two of his own workmen, he observed 'how very foolish it is for a day labourer to leave his station as a servant and ... become a little master'. Even those who remained labourers were getting higher wages and acquiring higher expectations, such as brick cottages instead of the mud and stud ones that had been acceptable in his own youth. He even regretted allowing his men to keep cows: 'So long as the head of a parish or a large farmer employs poor men with large families

---

[32] DIXON 7/1/11: 16 September 1813; 7/1/13: April 1816; 7/1/21: 23 June 1824.

[33] DIXON 7/1/5: 26 July 1807. William's papers have examples of accounts kept and even letters written by his workpeople (e.g. John Towers n.d., asking if his old father could have 'a letel employ at ome', DIXON 4/3, fo. 19).

[34] Cornelius Stovin was similarly hurt when his men joined the agricultural labourers' union (*Journals of a Methodist Farmer*, 62).

who have no cattle, i.e. cow or pig, so long does he keep his position and consequence.'[35]

* * *

William Dixon had little truck with the concept of class, but he was attracted to the idea of community. He used the word several times in his musings on local society, and on two occasions attempted to define it. 'Community', he wrote around 1800 in a little 'Callender for myself', 'means a partnership, the having things in common, a society of men inhabiting the same place (parish) or a body of men uniting in civil society for mutual advantage.'[36] This is somewhat elastic, but it embraces the notion of a parish as one species of local community. In March 1806, in the context of an economic argument about the local generation and distribution of wealth, he broadened the scope of his definition.

> If the produce of labour be taken from or out of a country or community where it is accumulated by its superiors it robs it of its rights and therefore impoverishes its members .... A community here is not meant a country but a parish or several parishes connected with the like markets, canals, merchants, prices and qualities.[37]

On another occasion he stretched the boundaries more vaguely still. 'The social principle in man', he opined, 'is of such an expansive nature that it cannot be confined within the circuit of a family or neighbourhood: it spreads into wider systems and draws men into wider communities.'[38] He understood that what defined a community was not just physical boundaries but the social bonds established within those boundaries. 'How necessary it is for Man's welfare', he wrote, taking his cue from Locke as well as the Bible, 'to become truly and sociably united.' And again: 'Man's condition cannot be bettered without a general concurrence in the great work.'[39]

In examining his sense of how these communities worked in practice one can do no better than begin with the parish. At one point he described it as an interlocking economic system dependent on the correct functioning of all its parts. It embraced not only the farmer but the craftsman, the tradesman and even the merchant. If one part of the system got out of kilter

---

[35] DIXON 7/1/2: 10 January 1806; 7/1/7: 16 February 1809; 7/1/8: 12 June 1811; 7/1/4: 24 November 1806.
[36] DIXON 7/3/2.
[37] DIXON 7/1/3: 14 March 1806.
[38] DIXON 7/1/15: 11 May 1819.
[39] DIXON 7/1/8: 25 May 1811; 7/1/15: 19 May 1819.

with another - if for instance wages ran ahead of prices – the equilibrium would be destroyed. In preserving this system the leading inhabitants of the parish had a crucial role. 'It is in the power of two men (persons) in a Parish the Landlord and Rector to make the rest of the persons or family in a parish rich and respectable or poor and despicable according as they are disposed to do good.' In the absence, however, of a resident squire or parson that responsibility would fall to the most substantial resident: the proprietor of a parish might not be 'of such consequence in common parish concerns as a respectable occupier'.[40]

In Holton this position fell to Dixon himself, and he was fully conscious of its importance, especially in the absence of a resident clergyman. 'I am particularly bound', he wrote in 1813, 'to persuade and even compel the several members in the parish to do their several duties with truth and justice, by shewing mercy to their weaknesses and enlightening their ignorance.'[41] Nearly everyone in Holton was dependent on or connected with him; and, like any dictatorial squire, he made it clear that they were expected to play their part in parish life. They were to attend church regularly,[42] and if they could sing well or play an instrument so much the better. The small tenants should attend vestry meetings – he saw the vestry as a parliament in miniature[43] – and be prepared to take the parochial offices. William Hewitt, the most substantial of the small tenants, certainly did his duty: he acted as parish clerk and sexton, served from time to time as constable and land tax collector, and was a useful oboe in the church band.

Beyond the parish, Caistor and its neighbourhood appears not infrequently in his meditations. He reflects, for instance, on Sabbath observance in the context of 'the customary usage of this vicinity of Caistor'. But he also gives various examples of *un*neighbourly behaviour. There was the wealthy Mrs Lawrence, for instance, who made use of the Caistor Canal Company's barrows but ordered her own coal from Brigg. Philip Skipworth earned a black mark as lord of the manor of Caistor by impounding the cattle that had strayed through his own failure to keep his fences in repair.[44]

---

[40] DIXON 7/1/8: 5 August 1811; 7/1/9: 24 July 1811; 7/1/7: 9 March 1809.
[41] DIXON 7/1/11: 2 September 1813.
[42] DIXON 7/1/7: 22 January 1809.
[43] DIXON 7/1/17: 23 May 1821.
[44] DIXON 7/1/3: 2 August 1806; 7/1/1: 26 February and 3 March 1805. Good neighbours, on the other hand, were a valuable commodity. When Francis Raynes did not take the south farm at Thornton in 1811 Thomas John wrote to express the Dixons'

In economic terms, however, Dixon often thought in terms not of a single market area but (as in the second definition quoted above) a district containing more than one market town but sharing the same or similar commodity prices, labour rates and credit arrangements. When in April 1816 he floated the idea of a credit scheme based on the price of wheat, he proposed that it should cover the wapentake of Yarborough, which included Barton and Brigg as well as Caistor.[45] It would be natural for him to think of the area over which his personal credit extended, and within which he was personally known and trusted. But it was above all the cost of labour that exercised him: 'I ought to be particularly careful and mindful in adhering to the average prices of my neighbourhood in purchasing labour.' In 1805 he refers to an unnamed local farmer who paid 9s 6d an acre for shearing wheat, as against the local rate of 5s 6d, and who had tried to entice away a neighbour's farm servant by offering him nine guineas a year. 'What is to be done with such members of the community?' William himself had been guilty of misguided charity during the dear times (probably of the mid-1790s), when he had kept ten labourers, supplying them with corn at a low price, and distributing rice on a Monday to those who had attended church on the Sunday. But that was before he had become 'enlightened' by reading books on political economy.[46]

Another type of community could be formed if a group of parishes came together for Poor Law purposes. Here he underlined the importance of neighbourliness, a virtue that was essential not just to good personal relations within a parish community but to good relations between parishes.

> Know that the word 'neighbour' means that man who will and doth show mercy to you tho' he dwells at a distance from you, and not that man that will injure you tho' he dwells near or adjoining your premises. Do not consider that parish a neighbouring parish who will dispute with you about the settlement of a pauper and expend more money in a lawsuit than would provide for the maintenance and support of the pauper during the whole term of years he had to live on this troublesome globe.[47]

---

disappointment, 'as good neighbours, what I am sure we should have found you, are particularly desirable' (DIXON 9/6/1).

[45] DIXON 7/1/13: 12 April 1816. The previous year he had blamed paper currency problems for the financial difficulties of the Caistor Society of Industry (DIXON 4/3, fo. 47).

[46] DIXON 7/1/1: 18 February and 26–27 March 1805. In 1778 his father had engaged one of his labourers to do some thrashing 'at a fair country's price' (DIXON 4/1, fo. 22).

[47] DIXON 7/1/1: 9 March 1805.

Settlement disputes could certainly embitter relations between parishes, as his father would have experienced as a magistrate. Both he and his father, moreover, had been overseers, and must have been well aware of the weaknesses of the Old Poor Law.[48] One way of avoiding friction was to create a local Poor Law union, and such a union could also become the basis of a charitable organisation extending over a number of parishes. In other words, a 'society of industry' could be formed as 'a general union of a neighbourhood'.[49]

* * *

It seems to have been in the winter of 1799–1800 that Dixon matured his plans for just such a 'general union' for the Caistor neighbourhood. Its aims would be to promote 'habits of Industry, Economy and Morality amongst the labouring Poor'. As for industry, he advocated a house or houses of industry for the spinning of yarn and parochial schools to teach spinning, weaving and carding. Industry and sobriety could also be encouraged by awarding premiums to religious and sober workmen.[50] Thrift was to be promoted through friendly societies and savings banks, while morality would be inculcated by Sunday schools and through the distribution of improving literature. The charitable efforts of the neighbourhood were to be directed both to rewarding the virtuous poor and to helping those in want through no fault of their own. The society that was to realise this programme was to consist of a voluntary association of parishes, but it would also rely on a body of personal subscribers.

None of the individual elements in this scheme was particularly novel. He had read about efforts in other counties in the *Farmers' Magazine* or in the reports of the Society for Bettering the Condition of the Poor. In Lincolnshire there had been a few local initiatives, usually coinciding with periods of distress and unemployment such as the early 1780s or the mid-1790s. In 1783 a Society of Industry had been formed at Spilsby, and had promoted the establishment of parochial spinning schools. Alford had a school of industry for a few years. In 1796 the Folkingham agricultural society was studying the rules of the Bath Agricultural Society as a possible model for giving premiums. William himself made use of the Bath society's rules, and the Revd Samuel Turner, Caistor's curate and a leading resident of the town, seems independently to have had the idea of

---

[48] Holton had had its own small poor house since 1784 (DIXON 11/6/5).
[49] DIXON 7/1/13: 28 July 1816.
[50] DIXON 7/6/2.

a Benevolent Society that would amongst other things distribute tracts.[51] William's scheme, however, was a good deal more ambitious than earlier initiatives in the county. His Society of Industry was to act as an umbrella organisation for a range of different activities, and it was, he hoped, to cover the wapentakes of Yarborough, Bradley Haverstoe and Walshcroft, an area stretching from roughly the Louth–Market Rasen road in the south to the Humber in the north and from the Ancholme in the west to the Humber coastline in the east. Caistor would make a good central point for this district, but the Society would be unlikely to succeed without a following wind from Brocklesby, and even then it might not gain many adherents in towns as far away as Barton or Grimsby.

To add further to the scope of the enterprise, however, it was to attach to these philanthropic objects what was essentially a Poor Law proposal, a scheme for a union of parishes with a central workhouse. This would enable the individual parish poor houses to be closed; and if the workhouse inmates were made to earn their keep, and its regime were made sufficiently unpleasant to deter all but the genuinely destitute, then there was every prospect of being able to reduce the poor rates. Again, these were not new ideas. A mechanism to enable parishes to form voluntary unions had been provided by Gilbert's Act in 1782, although it was not much used until the late 1790s.[52] In Lincolnshire only Lincoln had formed a union by 1800, and that, in 1797, was under a local Act rather than Gilbert's Act. But in 1799 there was a movement at Keelby, a populous semi-open parish five miles north-east of Caistor, to provide a workhouse for that and some adjacent parishes.[53] This came to nothing, but it indicated local concern over rising poor rates, and may well have encouraged the Caistor proceedings a few months later.

On 15 March 1800 Dixon convened a meeting at Caistor (to coincide with the Saturday market) at which his proposals were discussed, and this led to a much larger public meeting at the George Inn ten days later. At that gathering rules for a society on the Bath model were adopted. It was also decided to take up the question of a union under the Act of 1782, and it was this part of the scheme that attracted most support. The parishes were consulted; further committee meetings took place towards the end of the year; and the union was formally constituted in April 1801. It had been

---

[51] Sir Frederick Morton Eden, Bt, *The State of the Poor* (London, 1797), ii.390, 399; *Stamford Mercury*, 21 October 1796; DIXON 7/6/2, p. 23.
[52] Sidney and Beatrice Webb, *English Poor Law History: Part I, The Old Poor Law* (London, 1927), 275.
[53] *Stamford Mercury*, 10 May 1799; *Keelby 1765–1831*, 25.

agreed to build the workhouse on Caistor Moor, a largely empty tract of unenclosed waste below but fairly near the town. The site was relatively easy to acquire – being common land its transfer required merely the consent of the lord of the manor and the holders of common rights – although William later regretted that it had not been built at Holton. A plain but imposing building, bearing the inscription 'Society of Industry established Anno Domini 1800', it received its first inmates in 1802.[54]

The Revd Samuel Turner was made president of the new society. A friend of Dixon's, and sympathetic to at least some of his ideas, he was also well connected locally: his brother John was Caistor's leading solicitor; his cousin, another Samuel Turner, had the chief medical practice in the town; he had various links with the farming community of the district; and, not least, his brother-in-law was George Tennyson. For a quarter of a century Turner filled the chair at a succession of public meetings at the George, where he also resided at one time.[55] There were five vice-presidents – the Revd William Wilkinson of Grasby (another Turner connection), Thomas Marris of Barton (Caistor's lord of the manor at this date), Thomas Lawrence (a leading Caistor gentleman),[56] William Richardson of Great Limber (a highly respected farmer and Yarborough tenant) and William Holgate of Keelby Grange (another Brocklesby tenant). Dixon, who was going to do most of the work, was made secretary and treasurer, and was also to become Visitor of the workhouse. A long list of those who attended the meeting of 25 March 1800 included other leading inhabitants of Caistor and the neighbourhood, among them George Tennyson ('Esq'), Philip Skipworth, Robert Parkinson and Richard Roadley. The list confirms that this was a middle-class movement, with support from townsfolk and from some of the leading farmers of the district. The patronage of the nobility and gentry, on the other hand, was noticeably absent.

---

[54] William Dixon, *Reports of the Several Institutions of the Society of Industry, established at Caistor A.D. 1800, for the better relief and employment of the poor, and to save the parish money* (Caistor, 1821), i.53ff; DIXON 7/6/2; 17/1/10; 7/1/4: 8 November 1806. In August 1803 'Mr Lundie' [William Lumby of Lincoln?] was paid five guineas for the plan and estimate (DIXON 17/3).

[55] Dinah Tyszka, 'Clergy, church and people 1790–1860', in Rex C. Russell, *Aspects of the History of Caistor 1790–1860* (Nettleton, 1992), 41. Dixon looked up to him as a man of superior education, and in 1806 hoped that he would become a St Paul, 'although a Saul at present' (DIXON 7/1/3: 23 March 1806).

[56] Thomas Lawrence (1759–1804) had married into the Wrays, who were Caistor business people, but his mother was a Maddison of Stainton-le-Vale. He himself sat on the Grand Jury at Lincoln and was referred to as Esquire. (DIXON 15/1, Kirkby Pedigrees, vol. 30.) The Mrs Lawrence referred to in this chapter was his widow.

Gilbert's Act specified a union area with a ten-mile radius from the central workhouse, which in this case included the Caistor market area, most of Market Rasen's, but part only of that of Brigg and no part of those of Barton or Grimsby. (So much for William's much larger original scheme.) Of the original twenty-two member parishes, twelve lay between Caistor and Market Rasen, reflecting the influence of Dixon himself, Skipworth and Tennyson. The other ten lay on the Wolds north-east of Caistor, but in this area membership was patchy. The adherence of Aylesby, Healing, Ravendale, Riby and Laceby may be attributable to Skipworth, Parkinson, Roadley and Dixon influence. But Limber and Keelby did not join at this date, although both did so a few years later. Although Richardson of Limber and Holgate of Keelby joined the Society their parishes did not join the union, and this was undoubtedly a reflection of the attitude of their landlord, Lord Yarborough. He regarded his home parish of Brocklesby as a kind of Poor Law peculiar – outside the system altogether – and he may well have been reluctant to give his blessing to a movement led by a crotchety farmer and a bibulous tory parson.[57] And where Yarborough led smaller landowners such Thorold of Cuxwold would follow. Dixon felt the blow: 'Lord Y, without your support the House of Industry will lose ground. I cannot support it against the farmers, etc for if they are to have their will they would pay no poor rates or taxes.'[58]

* * *

Over the next few years the union, with the House of Industry at its centre, became part of the administrative landscape of the Caistor neighbourhood, enhancing both the importance of the town and Dixon's own local standing.[59] More parishes in the Market Rasen area joined, and in 1806–9 a few parishes between Caistor and Brigg, including Searby, were recruited. By 1810 most parishes within five miles of Caistor were in the union.[60] The manufacture of woollen yarn was begun in 1803, and by 1811 this was making £775 12s. But a farm in Nettleton attached to the House was less successful, producing a surplus of only £43 7s ½d in 1811.[61] William took

[57] Rawding, *Lincolnshire Wolds*, 207.
[58] DIXON 7/1/2 (2 January 1806).
[59] Society and House of Industry matters often feature in his accounts with his neighbours (DIXON 4/3).
[60] DIXON 17/1/1. The most far-flung parish to join was West Torrington, near Wragby, one of the parishes of which Samuel Turner was non-resident incumbent.
[61] DIXON 17/1/3; 7/6/4; 4/3, fos 227, 231v; Abstract of Returns relative to the expense and maintenance of the Poor (*Parliamentary Papers*, 1803–4 xiii), quoted by Rex C. Russell, 'Caistor Hospital' (unpublished typescript, 1973).

his position as Visitor very seriously, and spent much time overseeing the workhouse and trying to 'improve' its inmates, efforts not helped when the governor fathered an illegitimate child in 1807. In 1806 William obtained a schoolmaster from the well-known house of industry at Kendal to teach the pauper children,[62] and also engaged a teacher of psalmody.[63] William himself gave occasional lectures or homilies to the boys, which must have added in a small way to the deterrent effect of the regime. (A chapel for the House of Industry, however, had to wait until 1865.)[64] These public-spirited activities came at no small charge to his own purse. By April 1815 the union owed him £2,754 13s 7d.[65]

In quieter or more prosperous times his neighbours were content to indulge his fancies as long as the union was saving them money, but when the rates went up the grumbling was not slow to begin. There were mutterings in 1805 and again in 1807–8.[66] In 1811 voices were raised in criticism of the victualling account, and there was talk of a stricter set of regulations, the feeling appearing to be that the regime was soft as well as inefficient.[67] (It was certainly less efficient from 1812, when Marmaduke Dixon took over as treasurer.) In 1816 an investigating committee consisting of three clergymen, Turner and Bowstead of Caistor and Holiwell of Irby, vindicated the management against charges of 'inconsiderate profusion, or other gross abuses', but recommended some economies and tighter supervision. This quelled the opposition for a while, but two years later William was seriously considering resigning.[68]

* * *

The Society of Industry, as distinct from the Poor Law union, had a somewhat shadowy existence. There was a general fund, to which a few gentry such as Sir Montague Cholmeley of Norton Place, Bishop Pretyman and

---

[62] DIXON 7/1/3: 26 March 1806.
[63] DIXON 7/1/17: 10 May 1821 (a reused Society of Industry account for 1812); 4/3, fo. 123.
[64] Russell, 'Caistor Hospital', 9.
[65] DIXON 4/3, fos 123, 243v.
[66] In 1807 'Mr C' [William Codd of Nettleton, a leading Eardley tenant] declared that the Society had been long enough under the care of one individual, causing Dixon to complain of Society business and officers being discussed at 'public ordinaries' instead of at Guardians' meetings (DIXON 7/1/5: 6 November 1807).
[67] Despite the provisions of Gilbert's Act the House of Industry seems to have been intended as a deterrent to the idle and dissolute as well as a refuge for the sick and infirm.
[68] For the 1816 committee report, see DIXON 7/1/14.

later his son William Edward Tomline contributed.[69] But the total subscribed was small, and a plan to increase the charitable fund connected with the House of Industry in 1820 appears to have failed.[70] The Society never gave premiums to industrious labourers, and a scheme of William's for 'parochial settlements' or spinning schools was unrealised, except possibly in Holton itself for a few years from 1817.[71]

In other areas, however, his efforts bore more fruit. A friendly society was launched in December 1805, taking over a moribund society that had been started as long ago as 1769, and by June 1806 it had forty-six members. The Revds Samuel Turner and Rowland Bowstead were the treasurers, and George Jackson, the landlord of the George, the clerk.[72] This time there was more aristocratic support: the subscribers included Lords Yarborough and Eardley as well as Cholmeley and the Bishop. The greatest generosity, however, came from William's mother and Mrs Parkinson's executors, who gave £50 each. By 1812 the society had assets of £410 12s.[73] But William's most important charitable work, apart from the Society of Industry itself, was connected with the Sunday school movement. The largest school, at Caistor itself, was started in 1807, and in 1813–14 he erected a building for it, at a cost of £200. Some of this was met by subscription, but he was left with a large deficit. In 1812 he helped to provide a school at Keelby, and in 1820 one at North Kelsey, another large village. Pupils were expected to attend for four hours on a Sunday morning and another four in the afternoon, a timetable perhaps more popular with their parents than with the scholars themselves. As at the school in Holton William wanted the children to learn to write as well as read, but the core curriculum was of course religious. Among William's papers is a 'Divine Song' entitled 'Heaven and Hell', which begins

>   There is beyond the sky
>   A heaven of joy and love
>   And holy children when they die
>   Go to that world above.
>
>   There is a dreadful hell
>   And everlasting pains

[69] DIXON 4/3, fo. 252. Dixon and Cholmeley got on well together, despite their difference in status.
[70] Dixon, *Reports*, i.451. Only one parish replied to his circular.
[71] DIXON 7/4/21.
[72] Bowstead had come to Caistor as master of the grammar school in 1803.
[73] Dixon, *Reports*, i.30ff; DIXON 7/6/3; 7/1/1: 9–10 March 1805; 7/1/11: 8 August 1813; 4/3, fo. 259.

> Where sinners must with devils dwell
> In darkness, fire and chains.[74]

Altogether more cheerful was the organisation that became known as the Caistor Matrons' Society. This sprung from an idea of William's for a 'family friendly society' in connection with the Society for the Promotion of Christian Knowledge, and it successfully linked the S.P.C.K. with the children of the House of Industry and the local Sunday schools. In 1809 it held its first anniversary: the children were examined by the lady subscribers, after which there was a colourful procession to Caistor church, a service with an anthem and a sermon, and a prize-giving ceremony at the George, the prizes of course taking the form of S.P.C.K. tracts. These annual events became part of the Caistor calendar, and were even reported in the county newspaper. They also acknowledged the importance of that element in the society of Caistor and its neighbourhood, the well-to-do wives and widows. The patroness in 1810 was Mrs Lawrence, followed by Mrs Tennyson, Mrs Roadley, Mrs Turner, Mrs Skipworth and (in 1816) Mrs Dixon.[75]

* * *

Dixon continued as Visitor until his death in 1824, but during his last few years his influence in Caistor shrank. He was less energetic than previously; his greatest ally, the Revd Samuel Turner, was in an alcoholic decline;[76] and there was something of a resurgence of the Yarborough interest in and about the town.[77] It was led by W.E. Tomline, who was now living at Riby, and who took the lead in, among other projects, the Caistor Savings Bank in 1818.[78] This had long been advocated by Dixon as one of the planks of the Society of Industry, and indeed he became a trustee, with Thomas John taking a more active role as one of the managing directors and Marmaduke as treasurer. But Lord Yarborough became the leading

---

[74] DIXON 4/3, fo. 123 (enc.).
[75] DIXON 17/3/1; 4/3, fos 38v, 143; Dixon, *Reports*, i.424; *Stamford Mercury*, 11 August 1809, 4 June 1819; Rex C. Russell, *A History of Schools and Education in Lindsey, Lincolnshire 1800–1902, Part 2, The 'miserable compromise' of the Sunday School* (Lincoln, 1965), 29–33.
[76] He lost the Caistor curacy in 1823.
[77] The revival of the whig (Yarborough) interest in north Lincolnshire was no doubt assisted by the ending of the war with France and the consequent weakening of Church and State patriotism. See also Olney, *Rural Society*, 146.
[78] In 1815 Tomline had chaired a Caistor meeting to form a new S.P.C.K. committee for the rural deaneries of Grimsby (i.e. the wapentake of Bradley Haverstoe), Yarborough and Walshcroft, with Lord Yarborough as president – a clear incursion into Dixon territory.

vice-president, supported by a raft of gentry from the Brigg and Rasen districts as well as the vicinity of Caistor itself.[79]

In his latter years William occupied himself partly by a venture into publication. In 1821 he produced a three-volume work on the Society of Industry at his own expense. A curious compilation bulked out with lengthy extracts from the *Homilies*, Blackstone and other sources, it nevertheless provides some useful information about the Society and its work. It was dedicated to his 'dear friend' Sir Montague Cholmeley.[80]

William's philanthropic work was continued by James Green, of all his sons the most in tune with his own thinking and social attitudes. It was James who took over the Friendly and Matron societies in 1822, and who succeeded his father as Visitor on the latter's death two years later. But he was a difficult character, more likely to make enemies than friends, and by the early 1830s the state of the workhouse was attracting much criticism.[81] But both the building and, in an enlarged form, the union survived into the era of the New Poor Law. The building, indeed, survived the New Poor Law itself, and remained in institutional use until the late twentieth century.[82]

Of the related organisations of the Society of Industry, the friendly society was in difficulties by 1823, and failed some time in the 1830s.[83] The savings bank was still going in the mid-nineteenth century: a ledger covering the years 1855–68 survives in the Dixon papers.[84] But the institution with the most persistent life in it turned out to be the Matron Society, at least partly because it combined an attractive annual children's festival with a role for the middle-class wives of the Caistor neighbourhood. It survived in its original form until 1884, and was revived in 1894 by the vicar of Caistor, Canon Westbrooke. But the First World War extinguished it again, and a further attempt to revive it, by the Revd T.G. Dixon in the early 1920s, was unsuccessful.

---

[79] The initiative may have owed something to the example of the recent movements in Louth, Lincoln and Stamford (*Stamford Mercury*, 19 September 1817; 9 January, 6 and 13 February 1818).

[80] For these *Reports*, see note 54 above. The volumes cost William £120 each. Cholmeley presented Dixon with a silver snuff box inscribed 'From a friend to a friend of the poor'. It passed to his daughter Mrs Skipworth, who used to dispense liquorice pieces from it to her grandchildren (DIXON 11/6/1).

[81] A public campaign was led by Sir Culling Eardley Smith, a militant Evangelical who had succeeded to Lord Eardley's properties in the Caistor neighbourhood in 1834.

[82] The original building was empty when I was allowed to look over it in 1991.

[83] DIXON 7/1/20; White's *Directory* of Lincolnshire (1842).

[84] DIXON 7/2.

1 'Holton House' (The Hall, Holton-le-Moor): sketch by Amelia Margaretta Dixon, 1852. Amelia (1833–1906), second daughter of Thomas John and Mary Ann Dixon, and later Mrs Jameson Dixon, succeeded to the Holton estate on the death of her sister Ann in 1893. (Lincolnshire Archives, 3 DIXON 8/3/5.)

2 The Hall, Holton-le-Moor in April 2017. (Nicholas Bennett.)

3(a) 'Old Manor House from the Kitchen Garden': watercolour sketch by Amelia Margaretta Dixon, *c.*1852. The old manor house near the Hall, used at this date to house the farm bailiff, was later rebuilt as two cottages. (Lincolnshire Archives, 3 DIXON 8/3/25.)

3(b) The park at Holton, looking north-east towards Stope Hill Farm and the Wolds: watercolour sketch by Amelia Margaretta Dixon. This part of the parish, previously moor and rabbit warren, had only recently been converted to park land. (Lincolnshire Archives, 3 DIXON 8/3/15.)

4(a) The Church of St Luke, Holton-le-Moor: watercolour sketch by Amelia Margaretta Dixon, 1852. A view of the church from the south-east, before the rebuilding of 1852–4. (Lincolnshire Archives, 3 DIXON 8/3/18.)

4(b) The Church of St Luke, Holton-le-Moor, from the south-east, in April 2017. The building of 1852–4 was extended by the Revd T. G. Dixon in 1925–6. (Nicholas Bennett.)

1756. Sheep put into Winter Quarters.
In the North Side. Shearlings — 87
and Two Shear Sheep — 21
In Saxby Close She Hoggs — 43
In the Little Slows She Hoggs — 21
Nov.br tho 4.th 89 Ews put to M.r Rainor Tup
90 Ews of y.e Highist Skins put to My
Fathers Great Tup & 86: put to the
Little Tup: put to y.e Tups in all — 267.
Single Gimbers put by y.e Tup — 102
Sent to Limbor. She Hoggs. + 22.
Sent to M.r Maurice Turnips. 80 He Hoggs.
In the Ings. She Hoggs — 40
In the Great Slows 20 She Hoggs and
17 He Hoggs & two Gimbers.

Hired Eliz:th Linsby from May Day
1757. to May Day 1758 . . . . . 1 0 0
Oct.r 4.th 14: For a pair of Shows — 0 2 6
for pattons mending 6. oct.r 9 d   0 0 6
part of a Handkerchief 0 0 7
pot 6.d for her Gown making & a pair Stay . 7: 0 7 6
Feb.y 10: p.d for a pair of Shoes: 2:7: ap.l 16: p.d for her
Shoes mending. 9.d   0 2 11

Hired Andrew Cook from
May Day 1757 to May Day 1758

5 Farming accounts of Thomas Dixon (1729–1798), 1756–8. This page, from the earliest account book in the Dixon collection, shows Thomas farming at West Firsby and borrowing his father's (William Dixon senior's) 'great tup' (large ram). (Lincolnshire Archives, DIXON 4/1, fo.3.)

A:D: 1756.

William Dixon Son of Thomas Dixon and Martha his Wife Baptized July the Seventeenth

A:D: 1757.

Rachel Dixon Daughter of Thomas Dixon and Martha his Wife Baptized September the 25th

A:D: 1759.

Thomas Dixon Son of Thomas Dixon and Martha His Wife Baptized June the Second

A:D: 1760.

Richard Dixon Son of Thomas Dixon and Martha His Wife Baptized the Fourteenth September

A:D: 1762.

Martha Dixon Daughter of Thomas Dixon and Martha His Wife Baptized the Twenty Third July

A:D: 1764.

Mary Dixon Daughter of Thomas Dixon and [Martha his] Wife Baptized the Fourth June

6 Thomas Dixon's memoranda in his account book of the baptisms of his older children, 1756–64. (Lincolnshire Archives, DIXON 4/1, fo.139.)

7(a) Holton Hall interior: Charlotte Roadley Dixon ill in bed – sketch by Amelia Margaretta Dixon 24 August 1854. Charlotte died, aged eighteen, on 22 September 1854. (Lincolnshire Archives, 3 DIXON 8/3/13.)

7(b) Holton Hall interior: preliminary sketch by Amelia Margaretta Dixon, probably of the morning room. The young lady at the piano is probably her elder sister Ann. (Lincolnshire Archives, 3 DIXON 8/3/14.)

8 Miniature portrait, probably of Richard Roadley Dixon (1830–1871), only surviving son of Thomas John and Mary Ann Dixon, at the time of his marriage in 1860. (Lincolnshire Archives, DIXON 21/2/17.)

9 Record of daily farming operations, 1824, kept for Thomas John Dixon by his bailiff Charles Slater. This volume, covering the Holton, Nettleton and Thornton farms, is the first in a remarkable series of farming day books ending in 1897. (Lincolnshire Archives, DIXON 5/1/1.)

10 Portrait of Thomas John Dixon (1785–1871), probably by A. Salomé, late 1850s. (A photograph in the Dixon collection, Lincolnshire Archives.)

11 Portrait of Mary Ann Dixon (1800–1885), wife of T. J. Dixon, by Benjamin Hudson, 1839–40. (A photograph in the Dixon collection, Lincolnshire Archives.)

1822
1 day geting in potatoes   1
2 days cutting & seting potatoes   2
3 day weeding corn   3   3
3 days & half stubing thistles   3½
2 days gathering twich   2
5 days & half shakeing manure   5½
1 day dresing wheat   1
1 day & half at the thrashing machine   1½
2 days dresing wheat   2
15 days half at seeds and hay   15
2 days hoeing turnips   2
half a day breaking salt   ½
10 days & half gathering & takeing away & rakeing corn   10½

12 Account of farm work kept by Robert Fanfield, of Holton, labourer, 1822. (Lincolnshire Archives, DIXON 5/4/8/9.)

> January 14 bill 7 weeks
> March 11 bill 8 weeks
> 1 Day Dressing stoup hill
> 1 Day mending tips
> 2 Days geten in sand spred
>   clay
> 1 Day kiling pigs
> 1 Day fothenin rabits
> 1 Day Dressing
> 1 Day spreding clay
> 1 Day getes in
> 3 Days at thoreuton
> 1 Day spreden clay
> half Day helping to Dress
> 1 Day geten in at top barn
> 5 days and alf ant plow
> 2 Days planting trees

13 Account of farm work kept by William Wilson, of Nettleton, farm servant, 1822. This and no. 12 are examples from a number of small account books kept by workers on the Dixon farms at this period. (Lincolnshire Archives, DIXON 5/4/8/11.)

TO BE
Sold by AUCTION,

By Mr. JOHN BENNETT,

At the GEORGE INN, in CASTOR, on SATURDAY the 22d Day of this Inftant MARCH, 1806,

SEVERAL SHARES

IN THE

Castor Canal Navigation,

In LOTS, comprifing £.100 each.

☞ Credit will be given, on approved Security, for one, two, or more Years.

BALL, PRINTER, BRIGG.

AT A PETIT SESSIONS, held at the JUSTICE ROOM, in BRIGG, in the Parts of Lindfey, in the County of Lincoln, on Thurfday the 27th Day of Auguft, 1807, before Valentine Grantham, D. D. and John Hildyard, CLERK, two of his Majefty's Juftices of the Peace for the faid Parts, JOHN GREEN, of Bifhop-Briggs, and ROBERT GREEN, of South Kelfey, were convicted, upon the Complaint of Juftice Jones, Collector of the Rates of the Caftor Navigable Canal, in Penalties amounting to the Sum of Twenty feven Pounds, for delivering in falfe Accounts of their refpective Ladings, being the fecond Offence committed by them againft the Caftor Navigable Canal Act.

And at the fame Petit Seffions, WILLIAM JOHNSTON, of Great Grimfby, Higgler, was alfo convicted, upon the Complaint of the faid Juftice Jones, for damaging the Banks of the faid Navigable Canal.

Notice is therefore hereby given

THAT any Perfon or Perfons in future offending in any Manner against the Provifions of the faid NAVIGABLE CANAL ACT, will be profecuted to the utmoft Rigour of the Law.

JUSTICE JONES, Collector of the Rates.

SOUTH KELSEY, Sep. 3, 1807.

Ball, Printer, Brigg.

14(a-b) Religious notes and reflections by William Dixon (1756–1824), 1807. From a series of notebooks or journals made up from scraps of paper – in this case printed notices relating to the Caistor Canal, in which the Dixon family had an interest. (Lincolnshire Archives, DIXON 7/1/5.)

15(a) Mrs Jameson Dixon seated in her bath chair, with her retainers. From an original photograph in the Dixon collection. (Lincolnshire Archives, 3 DIXON 8/7/4.)

15(b) The bath chair in no. 15(a). Purchased by Mrs M. A. Dixon from J. Ward of 246 Tottenham Court Road, London in March 1871. It was acquired by the Revd T. G. Dixon from Mrs Jameson Dixon's executors in 1906, and was used by his daughters to take them for rides, pulled by a donkey obtained from Cleethorpes. (North Lincolnshire Museums Service, Normanby Hall Collections. Photograph by Nicholas Bennett.)

16 Double portrait of Charlotte Roadley Dixon and Amelia Margaretta Dixon by Benjamin Hudson, 1844. (Original in family possession. Photograph by Nicholas Bennett.)

6

THE MAN OF BUSINESS: THOMAS JOHN DIXON (1785–1871), THE EARLY YEARS

In Chapter 4 the Holton farms in the first quarter of the nineteenth century were shown to be a family business, with William Dixon in effect the senior partner, at least initially, but with his two farming sons, Thomas John (or Tom) and James Green as juniors. We described how Thomas John, even before leaving school at sixteen, began to assist his father, and how from 1805 to 1812 he farmed at Nettleton and Thornton, in partnership first with his uncle by marriage Richard Roadley and then with his father. This chapter traces his subsequent career from around 1810, when he reached the age of twenty-five, to the mid-1840s, a period during which he progressed from partner in the Holton farms to their sole proprietor, and became not only one of north Lincolnshire's leading agriculturists but a major landowner in his locality. The two key events in this story were his father's death in 1824 and his marriage in 1827. But his progress towards wealth and success was neither even nor inevitable. There were times when he might, in vulgar parlance, have come unstuck.

* * *

As a boy Thomas John could indulge on occasion in coltish high spirits with fellow scholars at Caistor Grammar School such as Brady Nicholson of Laceby.[1] But from early on he showed an energetic aptitude for farming and a remarkably hard head for business. Perhaps in this he was a throwback to his great-grandfather: it certainly marked him out from his father and his brother James. An important influence was undoubtedly his uncle by marriage Richard Roadley, a successful and intelligent landowning farmer with whom he got on well. Here is an exchange of letters dating from February 1807:

---

[1] DIXON 9/6/4/6, 10, 11; S.W. Nicholson and Betty Boyden, *The Middling Sort: The Story of a Lincolnshire Family 1730–1990* (printed Nottingham, 1991).

My dear Tom [wrote Roadley], If the frost continues I would not have you send the horses to Searby for ploughing, but shall be extremely oblig'd if you will send one horse that will draw well in the shafts for the use of leading manures as we are quite fast for want of one …. I was at Searby yesterday and such a set of lazy geniouses I never saw ….

Tom replied:

Dear Uncle, We will send you 4 horses to Searby on Monday, if the frost does not stop plowg this week. I hope you have taken the keeping you talked of we shall want for about 8 or 9 score sheep in a fortnight's time.[2]

Soon after this Roadley withdrew from the partnership to concentrate on Searby, leaving Thomas John to develop the farming methods at Thornton and Nettleton that became his hallmark in later years.[3] He kept a large flock, often taking additional fields for winter feeding, but he also extended his arable, from 146 acres in 1805 to 368 in 1812. This involved paring and burning the poorer land in Nettleton, and he is said to have ploughed farther up the steep Wold-side than any farmer before or since. His corn was sent mainly to Brigg from Riverhead (the Moortown terminus of the Caistor Canal) or sold to local customers, and his wool went to Yorkshire as was customary. But his marketing generally was much more enterprising than his father's. He attended the principal north Lindsey fairs and markets, and his autumn journeys could take him to Kirton-in-Lindsey, Horncastle or even Beverley. He sold livestock on commission at Wakefield, or occasionally had it driven down to Smithfield. At one point he became interested in wolds, or welds, a crop grown for its yellow dye, for which there were markets in London and Leeds. He kept his ear to the ground, and seems to have corresponded with merchants outside the county as well as reading the newspaper market intelligence.[4]

He must have had reliable foremen, because he was frequently away from home. Apart from commercial trips he indulged in a little hunting (the first and last Dixon to do so), and as early as 1802 he had joined the Market Rasen troop of volunteer yeomanry. This provided congenial male companionship, and he committed himself to the amateur military life with characteristic thoroughness, being promoted Cornet in 1808, Lieutenant in 1813 'without a dissenting voice' and Captain, in succession to Ayscough Boucherett of Willingham, in 1817. In 1807 he even considered joining

---

[2] DIXON 9/6/1; 9/6/4/1–2.
[3] For the Nettleton and Thornton partnerships see also Chapter 4 above.
[4] For his farming 1805–12, see especially DIXON 4/12; 9/1/5–12; 5/6/1–2.

the regular army: it helped a young man to know himself, he considered, 'and he sees a deal of the world'. His friends and family dissuaded him, but he did set out to see as much of the world as he could reach from Holton on horseback. In 1808 he made a trip to Norfolk, where he admired Thomas Coke's large-scale farming and his vast plantations – a model use for light sandy soil that may have influenced his later planting at Holton.[5] Two years later he made the first of a number of trips in the opposite direction. His uncle Robert Parkinson had inherited two farms in Herefordshire, at Tyberton and Madley, from his mother, old Mrs Parkinson of Caistor.[6] Tom accompanied him to supervise the properties, and later went alone when Parkinson could no longer manage the journey. On Parkinson's death in 1822 the farms came to Tom himself.

By 1810 Tom, now in his mid-twenties, might have been expected to find himself a wife. There had been hints of a friendship with Elizabeth Nicolson, Brady's sister. (The family were now living at Keelby.) But that lady described herself as a Methodist, and probably considered Tom too worldly, despite his assertion that he preferred sense in a woman to gold. 'Oh *money, money*', she exclaimed in one letter, 'what will not people *sacrifice* for thee!'[7] In fact neither Tom nor his brother James were to marry until after their father died.

* * *

Meanwhile, living cheaply at home, Tom was beginning to accumulate some capital, and he soon showed an interest in purchasing land, especially land that might benefit his farming business. These acquisitions are worth recording in a little detail because they mark not only his own rise as a landowner but also the eclipse of three old Holton families. His first, modest purchase was a cottage and croft in Nettleton in 1810, but then, in 1814, arose an opportunity in Holton. Benjamin Broughton's property came up for sale, and William, having no spare money himself, must have agreed that Tom should secure it if possible. The story of how he did so has come down to us. Poor Broughton was in the debtors' prison in Lincoln Castle, and Tom was driven to Caenby Corner to pick up the Lincoln coach at an early hour. In the coach he met a man who he discovered was on the same errand, but he stole a march on him by calling on Broughton

---

[5] DIXON 9/1: T.J. Dixon's diaries *passim*; 9/6/4/16a. Much nearer to home, of course, though on chalk rather than sand, were the vast Brocklesby plantations.

[6] These properties came from her mother's family, the Greens.

[7] DIXON 9/6/4/16. Elizabeth married Thomas Skipworth of Cabourne. Brady moved to Wootton, on the Brocklesby estate, around 1815.

and closing the deal while his rival was still at breakfast. Part of the agreement was that Broughton should live rent-free in a cottage at Holton for the remainder of his days, a promise that was kept by 'putting him into part of a blacksmith's cottage which was very small and had one small window and a door'.[8] This story, even if exaggerated in the telling, is of interest for Thomas John's reputation, and if true was a sad end to the long association between the two families. The price of £1,100 was raised with the help of a loan from Marmaduke.

Broughton's failure was followed in 1817 by that of a more substantial Holton farmer, Robert Bett. Bett owned four cottages and nine acres (previously Broughton property) in the parish, plus 43 acres on the Marsh, and rented the last Bestoe property in Holton, a holding of 170 acres. He had added to these commitments a farm at Ashby-de-la-Launde in Kesteven, and it was this over-extension of his business that seemingly caused his downfall when prices tumbled at the end of the Napoleonic Wars. Thomas John had the unpleasant duty of accompanying his father's old neighbour to the Fleet Prison in London, and then of supervising the disposal of his assets. He himself bought the Holton property for £700 in December 1817, and 15 acres at Theddlethorpe, which had failed to sell at auction, for £500 the following year.[9] By this time the last Bestoe property in Holton had also fallen into Thomas John's lap. It was held by two co-heirs, and in March 1816 he had declined an offer of one of the moieties for £2,500, declaring that 'the depression of the times and scarcity of money are much against the sale of land', and pointing out that the property was subject to full tithe and a seven-year lease. But he obtained the lease himself in 1817, following Bett's failure, and was then able to secure the freehold of the whole farm for £4,700. He paid half the purchase price in 1818–19, but the remainder not until 1825.[10] These purchases in Holton (for which see also Map 2) were followed closely by another in Nettleton, where in 1818 he agreed to pay £3,100 for a cottage and orchard in the village, with 53 acres of wold land allotted at the enclosure. This was not at the time very close to any other Dixon properties, although forty years later it was to become part of Prospect Farm. Again he spread the payment over five years.[11]

---

[8] David Maddison, Dixon's coachman, told the story to John Brown, who repeated it to T.G. Dixon around 1910. In T.G.D.'s time the cottage was used as a tool house by the schoolboys: it was standing empty in 1984, but no longer exists.

[9] DIXON 1/A/5; 1/B/4/10; 4/9, fos 94v, 234; 9/1/7.

[10] DIXON 1/A/6; 4/9, fo. 165; 9/6/1.

[11] DIXON 1/C/3–5; 4/9, fo. 25; 4/10, fo. 181.

He was able to use most of these acquisitions to enhance his farming business, both in Holton and on the Marsh; and when he finally took over Hall Farm from his father in 1823 his total holding covered 1,400 acres. In the less favourable economic climate of the time he could nevertheless benefit from economies of scale, moving his men round the farms as they were needed. But he did not cut back his expenditure. On the contrary he increased his outlay on under-draining, and by 1820 was spreading a thousand bushels of bones a year on his fields. He also went in for machinery in a large way, buying bone mills and thrashing machines not only for his own use but also to hire out to his neighbours. The thrashing machine circulated very locally, but the bone mill travelled to the large farms of a wider neighbourhood, including those of Messrs Sowerby and Coates at Beelsby, J.R. Atkinson at Binbrook and Matthew Maw at Cleatham.[12]

This was not the full extent of his entrepreneurial ventures in the early 1820s. He began business as a maltster in Caistor in 1820, buying barley from such leading growers as William Torr of Riby and George Whitworth of Acre House, Claxby, and sending his malt to the industrial markets of Yorkshire, Nottinghamshire, Derbyshire and Lancashire, his sales in 1822 totalling £1,140. He acquired a tan yard in Caistor at much the same time, obtaining hides from Hull and elsewhere and sending the finished products to Gainsborough and Halifax. Marmaduke was a partner in these ventures, as he was in a smaller hop-merchanting project, in which Richard Roadley and Samuel Stothard of South Kelsey were the other two partners.[13]

During these busy years Thomas John was still able to indulge in a little recreation. September, a comparatively quiet month on the farms, might see him at Cleethorpes or Mablethorpe for a few days. In 1818, having visited the Herefordshire farms, he went on to Gloucester and then 'through a fine manufacturing country' to Bath before spending a little time on the Isle of Wight.[14] But he was becoming increasingly in demand in his own neighbourhood. For a number of years he acted as constable and highways surveyor for Thornton – the latter post enabling him to ensure that there was a minimum of pot-holes between his farms and the river Ancholme at Brandy Wharf – and he also acted as agent for the non-resident rector of Thornton (until 1809 the Revd David Field of Laceby), collecting the

---

[12] DIXON 4/9; 4/10, fo. 227; 6/6/2; 9/1/14. Atkinson and Maw were both connections of Marmaduke's wife.
[13] DIXON 4/9, fos 240–61 *passim*; DIXON 16/3: diary of Dr. Parkinson, 7 April 1823. When put up for sale in 1834 the tan yard had 67 bark pits, 6 lime pits, other equipment and an 'excellent run of water' (DIXON 9/2/3).
[14] DIXON 9/1/18.

tithes and arranging for the Sunday duties.[15] In 1821 he was made a commissioner for taxes, for the Ancholme Drainage and for the local Court of Sewers. In 1821, also, Thomas Towler, a Yorkshire wool merchant with whom the Dixons had dealt for some years, went bankrupt, and Thomas John was appointed by the creditors to sort out his affairs, a duty that involved several trips into Yorkshire. Energetic as he still was, though approaching his fortieth year, he could not be in more than one place at once, and in 1823 he took on a bailiff at Holton to oversee the men and keep the routine accounts, an appointment certainly justified by the size of the business.

* * *

William Dixon died in December 1824. During his last few years he had given careful thought to his testamentary arrangements, wanting to be fair to all his children and, presumably, to ensure that none of them would feel the traumatic disappointment that his own father's will had caused him. In the second at least of these aims he was spectacularly unsuccessful.

As already described, he had given his daughter £5,000 on her marriage. His wife had her jointure estate, and she had also been left an annuity of £145, payable out of the Herefordshire properties, by her mother. There would therefore be no need to make the kind of provision for her that his father had made for *his* mother in 1798. The challenge was how to divide his estate equally among his three sons. In 1816 William had considered a simple division of his land into three equal portions: even his wife's jointure lands were to be included (put into hotchpot) if Thomas John as their eventual heir were agreeable.[16] But this three-way split would have disrupted the businesses of the two farming brothers. In 1821–2, therefore, William decided on a division of the land into two estates following its current division for farming purposes, with adjustments to equalise their value and a sum in cash for Marmaduke. In September 1821 Francis Raynes came over from Hatfield to conduct a valuation. Following his recommendations James Green would receive a total of 613 acres, mainly his Mount Pleasant and Gravel Hill farms, while Thomas John would receive the remainder of the unsettled estate in Holton and Nettleton (including Home Farm and the Moor, although he had not yet taken them over), a total of about 950 acres. James's allotment, although the smaller of the two, was the more valuable, at £14,500 as against £12,500 (both figures

---

[15] DIXON 4/9, fo. 185; 4/10, fo. 170.
[16] DIXON 3/1/6.

representing an executorship rather than a full commercial valuation). Marmaduke was to receive £5,000 in cash, but, to equal out his brothers' shares, £3,500 was to come from James and only £1,500 from Tom. Even so Marmaduke appeared on these figures to be the loser, but William was taking into account the fact that money had been spent on his professional training and establishment in business, whereas the farming brothers had had to borrow money to take on their holdings. Furthermore, William provided that any shortfall in his residuary estate should be met by James and Tom, in the same proportion as their contributions to Marmaduke's legacy. Perhaps anticipating such a shortfall William's other bequests were modest – £300 to his wife, 20 guineas to his daughter, £100 each to the Caistor Friendly Society and Caistor Sunday School, and £50 to Holton Sunday School.[17]

Trouble broke out as soon as the will was read. Marmaduke felt himself poorly treated, but Thomas John was more deeply aggrieved, regarding the will as 'so very much in favour of J.G.D. and considered so by all parties'. He, Tom, had spent a lot of money consolidating the Holton estate, only to find its best farm, Mount Pleasant, left away from it. The three brothers held a conference at Holton, at which Marmaduke allied himself with Tom in expressing 'great dis-satisfaction'. They prevailed on James, who was in need of ready money to meet his obligations under the will, to sell Mount Farm to Tom for £15,000, nearly a full commercial valuation, with £500 being 'given again'. Out of this James would add £500 to Marmaduke's legacy.[18]

Far from settling the matter, this arrangement left James with a long-lasting sense of grievance. William had been owed money, including £2,165 by the Society of Industry, but he also left debts amounting to £6,719, according to a calculation made in 1828, including £2,000 owed to George Skipworth, £1,700 to Robert Parkinson's heirs, £1,000 to Richard Roadley's heirs and £500 to John Ferraby of Owmby (by Searby). The deficit was around £3,500, of which under the terms of the will James was liable for £2,450. James's grumbles surfaced periodically over the following years. In May 1832 there was an attempt to draw a line under the matter, with Thomas John paying a final instalment of £500 for Mount Pleasant, but further errors in the executorship account came to light in 1834, by which date James had paid claims against the estate amounting to

---

[17] DIXON 3/1/4; 6/2/19.
[18] DIXON 2/1/6; 3/1/71; 3/2/8–12. The full price of Mount Pleasant represented about twenty-eight years' purchase.

£6,000. In December of that year an arbitration was conducted by George Marris, the Caistor lawyer, and John Wilkinson, the Barton accountant, which concluded that James's charge against Thomas John was 'entirely groundless'. In 1837 the whole question was re-examined by Wilkinson, William Torr of Riby and William Skipworth of South Kelsey, who again exonerated Thomas John, while conceding that 'from the tenor of their father's will James Green Dixon has been called on to pay considerably more money to make up the deficiency in their father's personal estate than his brother Thomas John Dixon'. There was an epilogue to this sorry saga as late as 1852–3, when James queried the precise terms of an obscure financial transaction between his brother and their father that had taken place in 1819.[19]

* * *

The year of Thomas Dixon's death, 1798, had seen a link established between the Dixon and Roadley families. The years 1825–7, immediately following William's death, saw that link greatly strengthened. In May 1825, even before William's will had been proved, Richard Roadley died after a short illness. Still only in his late fifties, he left a substantial landed estate and a large farming business. Most of the Searby estate, of about 1,420 acres, was in hand, while the Messingham estate of some 560 acres was mainly let to tenants. He left a widow (Tom Dixon's aunt Ann), a son of only seventeen, and two unmarried daughters, Mary Ann, aged twenty-five, and Charlotte, aged twenty-one. Mrs Roadley naturally looked to her eldest and most competent nephew for help in this family crisis, which deepened when in April 1826 young Richard Dixon Roadley died in London. Mary Ann, who had been nursing him there, retired exhausted to Harrogate, where she slowly recovered over the next few months.

Thomas John took his responsibilities as his uncle's executor very seriously, keeping the Searby farm and estate going and sorting out the legal business with the help of Marmaduke.[20] But after young Richard's death he also took on a kind of guardianship role in respect of the two fatherless girls. He had been seeing more of his elder cousin Mary Ann in recent years, and following his father's death it is likely that he had begun to think seriously about getting married. His mother encouraged him. 'I think

---

[19] DIXON 3/1/13, 36; 3/2 *passim*. Wilkinson had worked for George Skipworth, but it was new in the Dixon family to employ an accountant.

[20] James Green was Thomas John's fellow executor, and the Searby crisis seems to have resulted in a temporary burying of the hatchet.

my mother is determined to drive me into a different way of life,' he wrote to Mary Ann in April 1826, 'for whether I am in health or sickness she will not give me her society without it is quite congenial to her own feelings.' Thrown increasingly together by the situation at Searby, the cousins came to an understanding, although they did not marry until July 1827.[21]

Mary Ann's pre-nuptial settlement put her husband in legal possession of the undivided moiety of the estate that had passed to her on her father's death, but it gave her a power of appointment, and should she not exercise that power her moiety would descend on her death to her children equally. Her father had also left her a sum of £2,000: the 1827 settlement merged this sum in the inheritance, but in compensation allowed her £200 a year in pin money. It was intended that the Roadley properties would be divided, with Searby going to Mary Ann and Messingham to Charlotte; but first they had to be valued, rather as the Holton farms had been a few years previously, and it was not until March 1829 that a deed of partition could be drawn up. Searby was valued at £62,459, compared with £21,913 for Messingham, so in order to equalise the inheritance Mary Ann's share was charged with a mortgage of £19,110 in favour of Charlotte. Mrs Roadley, moreover, was to have £800 a year for her life, resulting in a charge of £400 each on her daughters' properties. Had Searby been let to tenants it might therefore have produced an annual income of only about £400 a year net of all charges; but it was Thomas John's intention to continue to run it himself as a large farm, adding it to his already extensive operations.[22]

In 1826 Thomas John had also been worried about Charlotte. She was being courted by Samuel Hall Egginton, the son of a Hull merchant, whom she had met at Harrogate. Thomas John hoped she would not 'lose her heart with a person who at present is a stranger to us all', and felt obliged to make some enquiries. What he learned alarmed him further ('I dare not name what I heard of Mr S.E.'), and he finally intervened to put a stop to the relationship.[23] In fact the couple did marry, but not until fourteen years later. By that time Egginton had inherited a very decent estate from his father, and his business in Hull was doing well. Thomas John, who dealt with the firm, could no longer claim that Egginton was a stranger or a fortune-hunter, and was won round sufficiently to act as a trustee for Charlotte's marriage settlement. But the Yorkshire gossip machine that had worked against Egginton now turned on Thomas John himself. In 1843,

[21] DIXON 9/1/21–5; 10/1.
[22] DIXON 2/2/6–8, 10; 3/7/27.
[23] DIXON 10/1/7.

sitting in the coffee room of Bainton's Vittoria Hotel on the water-front at Hull, on his way home from a visit to Harrogate, he overheard his character being 'brought into question by two strange gentlemen'. Described as 'a very large farmer, a most extensive agriculturist', he was alleged to have demanded that the Eggintons should settle £10,000 on Charlotte, thus pricing her out of the marriage market with the aim of eventually securing her property for his side of the family. The two strange gentlemen must have been disconcerted when Dixon stepped forward to identify himself and secure an apology.[24]

* * *

Although their father's will caused trouble among the three brothers, it did not lead to any complete breach. There were reasons why they needed to get on with each other, since the businesses that they continued to pursue overlapped at various points. In the case of James Green, however, his father's death marked a significant realignment of his affairs, although they had already begun to shift in the early 1820s. In 1822 Robert Parkinson had left him a farm in Rothwell, a wold parish about three miles southeast of Caistor. In 1824 he purchased an adjacent farm from John Walter Dudding (a Roadley connection through the Jackson family); and in 1825, using part of the Mount Pleasant money, he bought a third Rothwell holding from his brother-in-law George Skipworth. The three properties were run together to form a good wold farm of 532 acres.[25]

It was in that same year, 1825, that he left Holton and set up house in Caistor. Marmaduke had also been left some property by Robert Parkinson, in the shape of his mother's old house in Caistor, the White House in the Market Place, and James rented it as his new family home, bringing his bride there in July 1825.[26] She was Elizabeth Dauber, the daughter of a well-to-do Brigg corn merchant: she brought him only £200, but she could expect £15,000 on her father's death. Caistor was within reach of James's farms at Rothwell, Holton and Thornton, but he was also planning to set up as a corn merchant from his house in the town. In this he may have been encouraged by his father-in-law, but the opportunity had already been spotted by Marmaduke, who in 1823 was reported to have advanced £3,000 for the purpose of carrying on a trade in corn with George and William Skipworth. James's core customers did indeed turn out to be from

---

[24] 3 DIXON 5/4.
[25] DIXON 1/E/3, 11; 12/1/2, fo. 161; 12/1/3, fos 71–2; 2 TGH 1/25/6.
[26] Mrs Parkinson was said to have been the last person in Caistor to use a Sedan chair.

within the family – Holton, South Kelsey and Searby were producing large quantities of corn at this period – but he also built up custom among a number of other farmers in an area roughly co-extensive with the Caistor union.[27]

Marmaduke was as enterprising and keen to make money as Thomas John, though unfortunately lacking the latter's business talents. Apart from the corn trade venture he was also involved in the Caistor malt kiln and tan yard; and in 1826 he joined Thomas John and George Skipworth in a larger speculation, forming a consortium to purchase 912 acres of Trent-side land at Althorpe in the Isle of Axholme. The idea was to improve the estate by warping (a system of fertilising it with silt from the river) and then re-sell it in lots. The initial outlay was £40,000, of which Skipworth furnished half and the Dixon brothers one quarter each. The scheme ran into difficulties when the land market slumped in the late 1820s, and eventually, after several years of trouble in administering the estate, Skipworth took the unsold residue into his own hands. The operation had involved evicting farmers from their holdings, creating a good deal of ill-feeling locally.[28]

Meanwhile Marmaduke, without a partner and assisted only by his faithful (if periodically exasperated) clerk William Moody, continued the ordinary business of a country attorney, drawing up deeds and wills, administering trusts and executorships, arranging mortgages and so on.[29] The work was as much financial as legal, and involved, or should have involved, the careful keeping of accounts, and a careful separation of his own and his clients' money. As with James's corn business, the principal clients were family. The affairs of the Dixons, Skipworths, Roadleys and Parkinsons were, as we have seen, far from simple, and involved some long-running executorships. Marmaduke's marriage also brought him some work. His wife was Susanna Atkinson, the daughter of a Hatfield solicitor, and they married in 1814. Her immediate family were Yorkshire people, but there were connections closer to home: Susanna's mother was

---

[27] DIXON 12/1/4; 12/1/6, fo. 161; 12/1/7, fo. 71; 16/3, p. 243.
[28] DIXON 4/10, fo. 137; 6/2/1; 9/1/25; 9/2/5/21; 22/4/4, fo. 190v; North East Lincolnshire Archives, Parkinson papers, 542/2/14; The National Archives, Chancery records, C 123/1/7. The original members of the syndicate had been the Revd Thomas Skipworth of Belton in the Isle of Axholme, his brother George of Moortown and George Tennyson. Thomas John eventually made about £3,000 out of the speculation.
[29] According to a survey of c.1990, single practitioners were more likely to play fast and loose with their clients' money than partners in larger firms (*Guardian*, 8 February 1991). After the failure of Ingelow and Co.'s bank in 1825 Caistor lacked a bank office, and Marmaduke seems to have taken on something of that role.

a Ravendale Parkinson, and of her brothers one became a large farmer at Binbrook and the other succeeded Richard Dixon as rector of Claxby. The Maws of Bigby and later Cleatham were also Atkinson connections. Marmaduke had dealings of course with other large farmers in the Caistor area, borrowing large sums of money from some of them, and he also built up a clientele among the less substantial inhabitants of Caistor itself and the surrounding villages.[30] The borrowings were partly to finance an ambitious programme of land purchases on his own account. Unlike his brothers he did not build up a compact estate but assembled a portfolio of landed investments over a wide area of north Lincolnshire. His main holdings, apart from his share in the Althorpe purchase, were in Caistor itself and at North Thoresby, near Louth, where he also acted as steward of the Yarborough manor. Altogether he accumulated several hundred acres.

All this went sour in the unfavourable economic climate of the late 1820s. By 1828 Marmaduke was experiencing cash-flow problems, and his health was suffering. But he was 'tenacious of interference', maintaining that his practice was worth £3,000 a year and that he was personally worth £40,000. When he died in London at the end of 1830 the truth came out. George Marris, another Caistor solicitor, took over the practice in partnership with Charles Smith of Market Rasen, but they found that they had bought a pup.[31] Marmaduke's executors, George Skipworth and Thomas John Dixon, discovered to their horror that he had left not a valuable estate but a ramshackle heap of bad debts and dud investments. They had both involved themselves deeply in his affairs, and faced heavy personal losses. Early in 1831 Thomas John was even contemplating selling the Holton estate, or mortgaging it to that local Croesus (and creditor of Marmaduke's) W.E. Tomline of Riby, for £20,000. Instead Messrs Eyre and Coverdale, the solicitors who had acted as the London agents for the practice, advised that the estate should be put into Chancery. This relieved the immediate pressure on the executors, but turned the affair from a short-term crisis into a long-term disaster. The assets could not be realised fast enough to stop the outstanding debts, including unpaid interest, from mounting year by year. When a dividend was declared in 1842 the secured creditors (mostly the wealthier ones) were owed £42,666 and the unsecured ones,

---

[30] Farmers such as William Borman of Irby, a Skipworth connection, sold their wheat to James and lent money to Marmaduke.

[31] George Marris was the son of a leading Yarborough tenant. He had married a Turner, and William Skipworth of South Kelsey was his brother-in-law. Charles Smith came from a leading Catholic farming family at West Rasen, and had been acting for Marmaduke at Market Rasen.

including many small Caistor tradesmen and other local people, the astronomical sum of £90,648. The secured creditors got some of their money back eventually but the unsecured ones very little. A further distribution in 1854 still left £112,639 unpaid, and small unclaimed sums remained in Chancery as late as 1908. George Skipworth lost as much as £18,700, and Thomas John over £11,000, not to mention time lost in attending meetings in London and elsewhere. Mrs Marmaduke lost her income, and her house with its elegant middle-class contents. (James Green eschewed the fancy furniture at the auction sale, contenting himself with sixty-eight bottles of wine from the cellar. Later he bought his own house from the estate at one of the property sales.) Mrs Marmaduke, who had no children, retreated to Hatfield, and later to Doncaster, where in 1859 she died of burns when her dress caught fire.[32]

For Thomas John the experience was a painful one, quite apart from the anxiety and pecuniary loss that it entailed. He had signed documents and stood surety for his brother to an extent that he would never have permitted himself in dealings outside the family, and in doing so he had contributed to a debacle whose consequences reverberated round the Caistor district and beyond. He had previously enjoyed, if that is the right word, the reputation of being 'a very artful man'.[33] Now people questioned his competence and judgment: 'T.J.D. they say ought to have found it out in late M.D.'s lifetime.' According to one later, and unattributed, source, he was never quite the same man again. M.D. himself was buried at Holton. Several decades later Mrs T.G. Dixon speculated that he had not died but had absconded to France to escape his creditors, leaving only a coffin-full of stones in the family vault.[34]

* * *

Thomas John's obligations towards Marmaduke's creditors were not his only heavy commitments in the decade 1825–35. By 1824, as we have seen, he had over 1,400 acres in his own occupation, and as a landowner he had acquired 590 acres in Holton, Nettleton, Thornton and Theddlethorpe, plus the Herefordshire farms. His father's death did not have a great effect on his farming acreage: he had already taken over Home Farm and the warren, and for the time being James continued as his tenant at Mount

---

[32] DIXON 9/2; TNA, C 33, C 123. For Susanna Dixon's death, see DIXON 15/1: Kirkby Pedigrees, vol. 23, p. 1.
[33] North East Lincolnshire Archives, Parkinson diary 1831, 242/2/14.
[34] DIXON 9/2/4/3; 11/6/1; information from the Misses Joan and Dora Gibbons.

Pleasant. But by 1825 he was the owner of 1,800 north Lincolnshire acres, and his marriage to Mary Ann Roadley increased his operations still further. By the late 1820s he was farming 1,500 acres based on Holton and a further 1,100 acres at Searby; and, counting the estate that he held in right of his wife, he controlled well over 3,000 acres in all. In 1828 he reckoned the value of his real property at £54,750, and his personalty, including farm stock and house contents, at £14,580.[35]

This impressive position had not been attained without incurring debts, partly in the form of mortgages but also in loans from family members and others. He owed around £10,000 by 1824, so it was fortunate that he could finance the purchase of Mount Pleasant with the sale of an outlying property. The Tyberton farm went to the squire of that parish, Daniel Lee Warner, for the very decent price of £10,500 in 1825.[36] Other, smaller, properties were disposed of in the next few years – the Theddlethorpe land in 1826, the other Herefordshire farm in 1834, and some of the Marsh property following his mother's death in 1842. Even so the Marmaduke crisis forced him to mortgage part of the Holton estate itself, in connection with a loan of £7,000 from James's father John Dauber in 1832. His liabilities continued to grow, and by 1835 they reached a peak of £25,000, a figure that included a debt of £6,000 to his mother and other sums relating to the still-running Dixon and Parkinson executorships.[37]

Remarkably these debts did not inhibit him from making two purchases that he deemed essential to the Holton estate. In 1831 he acquired the lease of the Holton tithes for £3,400, having obtained the first refusal several years before.[38] And in 1840, nearly two hundred years after it had been alienated from the Bestoe estate, he was able to purchase Ewefield following the death of its nonagenarian owner Mrs Shore.[39] The price was considerable – £6,800 – but by 1840 his affairs were luckily in a somewhat more prosperous state.

Luckily too the Searby estate and farm had been well managed by Richard Roadley, and Thomas John did not have to invest heavily in further improvements. The farm was large but not unwieldy (the Marshalls' farm at nearby Elsham was a similar size), and it was more productive, acre for

---

[35] DIXON 6/1/2.
[36] DIXON 4/10, fo. 57; Herefordshire Record Office, Lee-Warner papers, K8/80.
[37] DIXON 1/E/11/7–11; 4/14/1; 4/9, fo. 31; 4/10, fo. 127.
[38] DIXON 1/E/12/1–2; 4/10, fo. 36; 11/1/10/2; 9/1/21 (December 1822).
[39] DIXON 1/A/2/38–9; 9/1/38. See also Map 2. Thomas dealt over this purchase with his old acquaintance Edward Humble of Renishaw, Mrs Shore's trustee and himself a former tenant of Ewefield. At Renishaw he acted as steward to the Sitwell estate.

acre, than Holton, containing a good balance of wold and low-lying land. By the early 1840s wheat had overtaken barley as the principal cash crop, and the farm was yielding an operating profit of around £2,000. Thomas John was responsible for meeting the heavy charges on the estate, but he was helped by the fact that in addition to the profits of the farm he could obtain a rental income from the land and cottages that were not in hand: they yielded £554 a year by 1845.[40]

He managed the Searby farm with the help only of a foreman during most of the period 1825–40. He did most of the more important buying and selling himself, buying in bulk where he could save by doing so and co-ordinating the sales with those from Holton. At Holton, however, he had by 1825 evolved a more sophisticated system of management, with experienced foremen at Holton, Stope Hill and Thornton and a bailiff at Holton to keep the accounts and co-ordinate the day-to-day work of the business as a whole. Charles Slater came as bailiff in 1824, at a salary of £35 a year, but his service ended tragically when he was found drowned in a 'lake' (presumably a clay-pit) on the Breamer in December 1832.[41] He was succeeded by John Dudding, a young man from Cuxwold who came of good north Lincolnshire farming stock.[42] He was paid £60, rising to £70 when he undertook some corn and stock sales from Searby as well as Holton, and must have given satisfaction, for when he left in 1838 Thomas John lent him £200 to help set him up in a good-sized wold farm at Thoresway.[43] His successor, Joseph Mountain, started at Dudding's final salary of £70.

Despite this increasing reliance on managers Thomas John retained control over the more strategic decisions at Holton, particularly as they related to his longer-term plans for the improvement of the farms. During these years he ploughed up the poorer pastures, under-draining them (a special plough being bought in 1826) and preparing them for crops of turnips and corn, operations that involved considerable expenditure in labour, drainage tiles and fertilisers.[44] Where the land was poorest he initiated

---

[40] DIXON 5/5 *passim*; 6/1/11. The estate had been increased to 1,450 acres by a small purchase in 1837. The original mortgage had been presumably been paid off by Richard Roadley's executors.

[41] DIXON 9/1/31. This was altogether a grim time at Holton.

[42] His branch, the Duddings of Sawcliffe in Roxby, farmed under the Elwes family (DIXON 15/1: Kirkby Pedigrees, vol. 33).

[43] DIXON 22/4/2, fo. 214; 22/4/4, fo. 48; 22/7/5/2. Dudding later farmed at Howsham Barff, the Kelk's old farm.

[44] Dixon applied liberal quantities of bones to his turnip crop. These came from Hull, as did oil-cake for feeding his cattle.

a programme of tree-planting in the mid-1820s that continued into the early 1840s. As the trees matured their value as timber increased; and they also provided wind-breaks, shelter for game and an enhancement of the appearance of the estate. The work was concentrated in the eastern half of the parish, and it was the Moor that was his main target. By 1838 the warren had been reduced in size; Holton had acquired 111 acres of plantation (with an additional 25 acres in Nettleton adjoining a similar project on the Eardley–Smith estate); and a carriage drive had been constructed from the Hall to the Market Rasen–Caistor road, providing better access to the Moor as well as a route for Mrs Dixon's outings in the carriage. But all these measures were mere preliminaries to the final assault on the Moor, which took place between 1838 and 1843. In an almost military operation the warren walls were (mostly) levelled,[45] the furze and bracken cleared away, the land under-drained, the sandy soil mixed with clay and lime, the new fields fenced in, the planting scheme completed, and many loads of manure carted on to the land in preparation for its first crops.[46] The work employed not only the regular Holton men and their teams but gangs from Nettleton and Caistor; and Michael Atkinson, the Holton brick maker, was kept busy making thousands of bricks and draining tiles. By 1843 the common moor had been obliterated, and the appearance of that side of the parish transformed.

Why was this final push delayed until 1838, and how were the legal obstacles overcome that had thwarted the enclosure of the Moor in William Dixon's time? By the late 1830s Thomas John's finances were healthy enough to enable him to contemplate such a large expenditure, and the economic climate was more favourable than it had been earlier in the decade. By the late 1830s, moreover, he owned the whole parish except for two farms. One of these was Daisy Hill, which may never have had common rights over the Moor. The other was Ewefield, where similar rights, although they had undoubtedly existed in former times, may not have been exercised for many years, and where he may already have secured an option on its purchase. As for the cottagers who used the Moor as rough grazing, they had no rights as such, and could be given alternative land elsewhere in the parish. But probably the key factor was connected with the Holton tithes. Thomas John was now their lessee, and under a tithe

---

[45] Remains of the old burrow walls, however, could still be traced inside the plantations in 1985.
[46] DIXON 4/10; 5/1/12–16; 22/4/2. In these years Thomas John obtained fertilisers from wherever he could – bark from the Caistor tannery, soot from the Market Rasen chimney sweeps, sprats from Grimsby, mussel shells from Cleethorpes and ashes from Hull.

award of 1838 they were converted to a rent-charge. The award described the remaining 410 acres of open ground merely as moor (as recently as 1834 a survey had called it common); and there was nobody to object if, once this land had been assessed as moor for tithe purposes, it was later converted to arable.[47]

* * *

Their honeymoon in 1827 took Thomas John and Mary Ann to Cheltenham and Herefordshire, but before that they passed some time in London, where they spent £771 on furniture and equipment for Holton Hall. Among the purchases were Spode and Copeland china and a Broadwood piano.[48] Once back at Holton Mary Ann instituted a regime somewhat more comfortable than her mother-in-law's, with a footman and a cook in addition to the two housemaids. The footman engaged in 1829 came from Colonel Elmhirst's at Usselby House, and would have been a cut above a groom, but the fact that his clothes included a gardening jacket as well as two morning jackets indicates the range of his duties.[49]

Mary Ann soon became a mother. Ann, born in 1828, was followed by William, who died in infancy the following year, Richard Roadley (1830), Amelia Margaretta (1833), Thomas John (1835), Charlotte Roadley (1836) and Mary Ann (1839). In the wake of the new arrivals came a nurse maid, later with a girl to assist her, and from the late 1830s a governess. The growing household required a larger house, and it was twice extended, in 1830 and again in 1839–40. The first alterations were modest, and designed principally to create more space. At the front of the house the stairs were extended to the attic storey, enabling nursery rooms to be created above the main bedrooms. At the back a new kitchen was added, with new back stairs near it, and it was probably at the same time that a passage was inserted down the middle of the house, from the front hall to the back quarters, differentiating the family part of the house more clearly from the service areas. The former kitchen and scullery became a breakfast room and an office or business room, the latter conveniently overlooking the yard.[50]

The arrival of the governess, and the need for a school room as well as a nursery, seems to have prompted the more extensive changes of 1839–40.

---

[47] Lincs. Archives, tithe award F101; Olney, 'Enclosure of Holton-le-Moor'.
[48] DIXON 4/10, fo. 200.
[49] DIXON 4/10, fos 101, 206.
[50] DIXON 4/10, fo. 112; 18/6/1/3–5 (drawings).

The front stairs were rebuilt in a new position behind the entrance hall, and a further extension was made to the back of the house. This enabled various rooms to be redesigned or given different functions. The old drawing room, to the left of the entrance hall, became the school room, and the dining room opposite it became a breakfast room. A new and larger dining room was placed behind the breakfast room, and a new drawing room created on the first floor. To help cover the new wall spaces thus created a series of family portraits was commissioned from the artist Benjamin Hudson, a north Lincolnshire man. These changes enhanced the impression made on a visitor to the house, who would enter by the more spacious hall and ascend the rather elegant stairs before reaching the drawing room with its Italian fireplace. Behind all this dignity the extension provided a laundry room, book room and servants' hall on the ground floor (although Thomas John took over the laundry room for his office), and a night nursery, bedroom (for the governess?) and dressing room on the chamber floor. The Dixons could now enjoy a bedroom with a shower bath and two dressing rooms off it; and water closets were installed on both the ground and the chamber floors. This was plumbing Harrogate-style rather than Caistor-style, and brought Holton into line with early Victorian national standards of comfort and hygiene. The changes of 1839–40, however, did not advance the house to the status of a country gentleman's seat. The modest front elevation was unaltered, and the new additions behind it were carried out in a rather dull brick from Nathaniel Tupling's yard at Wrawby. The work was done by a local builder, Richard Colton of Moortown, and the bill came to only £600 – although that was more than the whole house had cost in 1785.[51]

The immediate surroundings of the house, meanwhile, were becoming less obviously agricultural. The rabbit house became a coach house;[52] the trees around the house were maturing;[53] and part of the warren and moor became parkland. There was no attempt, however, to give Holton the appearance of an estate village. Utilitarian farm buildings were erected as necessary – a large barn for the home farm in 1829, and later additional buildings at Stope Hill – but labourers' cottages were not high on the agenda: Thomas John lodged his men in redundant farm houses where he

---

[51] DIXON 4/14/2 enc; 18/6/1/8–9; 22/9/2, 17.
[52] DIXON 22/9/2/9.
[53] Mary Ann's axiom for picturesque planting was never to place more than two trees in a straight line (information of the Misses Gibbons).

could.[54] Nevertheless, although Holton did not look like a closed village and parish, the influence of the Dixon family had strengthened since William's time. Thomas John was the only substantial resident in the place. There was nobody of independent means, and Edward Brown of Daisy Hill, at 160 acres the only moderate-sized occupier, although taking his turn as overseer, seems otherwise not to have played much part in parish life. Trade was represented only by the estate carpenter and blacksmith. Three small tenants survived into the 1830s, Hewitt, Noble and Maddison, but Thomas John took no trouble to foster the dying breed of cow-keeping cottagers.

Like his father before him Thomas John assumed the principal responsibility for the institutions of the parish, principally the church and the Sunday school. When he entered his pew for the Sunday service he 'stood up and turned to the west [and] prayed into his top hat as the custom was. He then looked round the church and noted who was present. Then he pushed his fingers through his hair making it stand up and sat down.'[55] Mrs Dixon supported her husband's interest in the village, patronising the Sunday school and organising Christmas clothing distributions, no doubt much as her mother had done at Searby.[56] On weekdays Mr Dixon would expect to be greeted deferentially by his tenants and workpeople. But in a parish of only a hundred and fifty inhabitants, with many of whom he came into regular contact, it was difficult to maintain social distances. Holton's blend of formality and informality is captured by an anecdote involving the 'squire' and one of his workmen. Passing him when he was balanced on the top of a ladder, Dixon called up to him to reprimand him for not tugging his forelock: 'John, you haven't made your 'beisance!'[57]

* * *

When she was not busy in the house or visiting in the village Mrs Dixon would be expected to pay calls on her relatives and neighbours. A new

---

[54] When Thomas John needed a new cottage at Stope Hill he had Brice's old cottage on the Breamer, apparently a lath and plaster affair, dismantled and re-erected on its new site (DIXON 22/9/2/1).

[55] As told to T.G. Dixon (DIXON 11/6/4).

[56] *Stamford Mercury*, 26 December 1834, 14 January 1841. I owe these references to Mr Rex Russell.

[57] Information from the Misses Gibbons. The John or Jon in question is said to have been Jonathan Mills, who later returned to Holton to work as a builder for the Revd T.G. Dixon. On another occasion 'old John Paddison', who disliked making his obeisance, was given a heavy job until he conceded the point ('Holton Gazette', 7 September 1915, in private possession).

gig was purchased for the purpose in 1830, but there seems to have been a consciousness that what passed muster in north Lincolnshire might not be smart enough for Harrogate. In 1836, only six years after the purchase of the gig, the Dixons acquired their first four-wheeled carriage, a phaeton from Walkinden of Louth that cost £33. At the same time John Rysdale or Ridsdale, he who had joined the family as footman in 1829, was elevated to the dignity of coachman, and all as a preliminary to a visit to Harrogate two months later. In 1845 Thomas John spent £163 14s on a superior phaeton, this time purchased in Harrogate itself.[58] The nearest place to visit from Holton was Moortown House, where old Mrs Dixon had moved on her son's marriage, but one suspects that Mary Ann did not find her sister-in-law and mother-in-law particularly congenial: closer socially were the William Skipworths at South Kelsey Hall. Closer still to her heart, though physically at a greater distance, were her mother and unmarried sister at Searby. Mary Ann, supported by her husband, took a personal interest in the Searby estate, and one of her projects there was to restore the little church in 1832–3 as a memorial to her father and brother. The results were architecturally disappointing, but the interior was beautified with a gallery and stained glass windows.[59]

Beyond the family circle those on whom she might call included George Marris, her husband's Caistor solicitor, Dr Porter of Caistor, the family doctor in succession to Dr Turner, and the Bowsteads. At the country house level the Tennyson d'Eyncourts of Bayons Manor were on calling terms, and Thomas John, though not his wife, would receive the occasional invitation to Brocklesby. Thomas himself was generally too busy for this kind of social life, but his business trips to London might include dinner with an acquaintance or a visit to the theatre, while he made time when he could to accompany the family on holidays to Cleethorpes and later to Harrogate.[60]

In the previous two Dixon generations the marriage of the head of the family and the subsequent arrival of children had increased the number of social connections locally. In the case of Thomas John and Mary Ann this did not happen to the same extent, partly because as cousins their circles overlapped anyhow, and partly because the Roadleys were not a prolific

---

[58] DIXON 4/10, fo. 230; 4/14/2, p. 23; 9/1/29; 22/4/4, fo. 158. Later in the century the phaetons were replaced by a brougham and, in 1895, a landau.

[59] DIXON 18/1/3/8; 22/9/5; 3 DIXON 5/9; *Stamford Mercury*, 20 April 1832.

[60] DIXON 9/1: T.J. Dixon's diaries, *passim*; DIXON 10/12, M.A. Dixon's journals and accounts. A house was taken at Harrogate in 1836. From 1840 the town could be reached by coach or carriage *and rail*. In 1844 the autumn visit to Harrogate was preceded by a trip to Scotland.

family. Within Thomas's own family his elderly mother continued to represent a link with the Parkinsons, but her brother Robert had had no children, and the Ravendale branch of the family was not close. Thomas himself seems to have been too busy to foster old friendships such as that with Brady Nicholson. Of his two brothers, Marmaduke died comparatively young, and his memory remained a painful one. James Green continued to live in Caistor, but he had his own family, his own business and his own role in Caistor's social life. (He held an annual dinner at the White House, his home in the market square, to coincide with the fair.[61]) Comings and goings between the White House and Holton Hall were not frequent.

Nevertheless, although William Dixon's mantle as regards his charitable work had fallen on James Green rather than Thomas John, the latter did not shirk the public duties that came his way. The Market Rasen Yeomanry was disbanded in 1828, but his service as a volunteer was recognised in 1831 when he was made a Deputy-Lieutenant for the county. He gave up parochial office-holding in the late 1820s, and in the 1830s delegated the collection of the Thornton tithes to his bailiff, but in 1836 he was made a guardian of the new Caistor Poor Law union; and other matters such as drainage or tithe commutation required his presence at Caistor from time to time. The 1830s were also a busy time for agricultural and political meetings: the two became even more closely connected when numbers of tenant farmers were enfranchised under the 1832 Reform Act.[62] Thomas John was elected a vice-president of the newly formed North Lincolnshire Agricultural Society in 1836, and also supported the cause of agricultural protection (that is, the maintenance of the Corn Laws). In party politics he was a Conservative by conviction, but locally he was a semi-detached adherent of the Brocklesby interest, splitting his vote at general elections between the whig or Liberal (Brocklesby) candidate and the moderate Conservative. He did not, however, take the chair at public meetings at Caistor, where that role tended to fall to his brother-in-law George Skipworth.

In 1837 the Lord Lieutenant of the county, the first Earl Brownlow, was working on a new Commission of the Peace. As on a similar occasion fifty years before, there was an acute shortage of suitable candidates for the county Bench, but 'suitable' now meant almost exclusively the gentry and superior clergy. Brownlow and his leading magistrates were having to consider 'a class of persons not heretofore thought of sufficient weight and influence in the county', but they were doing so with great

---

[61] DIXON 12/5: pocket books of J.G. Dixon.
[62] Olney, *Lincolnshire Politics*, esp. 94–100.

reluctance. George Skipworth took the initiative locally in mentioning to Lord Worsley, Lord Yarborough's son and one of the M.P.s for North Lincolnshire, that both he and Dixon would be willing to serve, and Worsley forwarded their names to Brownlow. There were no petty sessions at Caistor at that date, and if appointed the two would sit at Market Rasen, where the Revd William Cooper was the leading J.P. Cooper learned, presumably from Brownlow, that these names were being canvassed, but he was unaware of the approach through Worsley, and thought that Charles Tennyson d'Eyncourt might have recommended them to the other sitting Member for the division, R.A. Christopher. Whatever the background, Cooper had considerable doubts as to the suitability of either Skipworth or Dixon. He described them to Brownlow as 'highly respectable and very opulent yeomen [or owner-occupying farmers], [who] have each perhaps an income of four or five thousand a year, and are Conservatives' (as of course were Cooper and Brownlow themselves).

> In 1835 they voted for Lord Worsley and Mr Corbett [i.e. they split their votes between the whig and the tory]; and tho' both are rather timid men would shew themselves, if a crisis for a decisive choice should arrive, I have little doubt, good and true. Their education has not fitted them completely for magisterial duties; but they would not disagreeably oppose their more experienced brethren.

If they were put on the Bench, Cooper concluded, it should be made clear to them that they were not beholden to the Liberal Tennyson d'Eyncourt for their elevation. In the event Skipworth was included in the commission, presumably as being the less 'timid' and more socially polished of the two. Dixon, still in 1837–8 more of the working businessman, was not.[63]

* * *

In 1837 Thomas John thought it prudent to make arrangements for his family in the event of his possible early death, just as his father had done in 1791. There were his wife and five surviving children to consider, and there were now two estates involved rather than one. This gave him more flexibility, but he needed to ensure that neither Searby nor Holton was disproportionately burdened with charges in favour of his widow or daughters. As far as Searby was concerned, it had already been settled that it would pass to his wife for her life, and since the birth of Richard Roadley

---

[63] Lincs. Archives, Brownlow papers, 4 BNL box 3; Olney, *Rural Society and County Government*, 98–101. In fact Skipworth's income cannot have been far short of £6,000 a year at this date, half of which was in rents from the entailed South Kelsey estate.

Dixon there seems never to have been any doubt that he would succeed to Searby on her death. In 1837 it was further decided that Thomas John junior, as the only other surviving son, would inherit Holton. The will that Thomas John made in that year left £6,000 to Mary Ann, £4,000 in trust for Richard, £6,000 for the maintenance of his three daughters, and all Thomas John's unsettled real estate, after these charges had been met, to young Thomas John. At the same time Mr and Mrs Dixon in a separate deed confirmed Richard as the heir to Searby but charged that estate with a further £12,000 in favour of their other children.[64] The combined effect of these two instruments was to provide total portions of £5,000 each for the three daughters. Their future was thus secured, but had Thomas John died soon after making this will a partial break-up of the Holton estate would undoubtedly have followed.

[64] DIXON 2/2/15; 3/4/1.

7

THE MAN OF PROPERTY: THOMAS JOHN DIXON, THE LATER YEARS

At the start of the 1840s Thomas John Dixon's financial affairs were still far from straightforward. He had long-term liabilities in connection with his brother Marmaduke's posthumous bankruptcy, and had recently committed himself to two major items of expenditure, the purchase of Ewefield and the improvement of Holton Moor. His debts, over £25,000 in 1835, still stood at nearly £23,000 seven years later.[1] But his assets were now very considerable. By 1842 his own landed estate, mainly in Holton, Nettleton and Thornton, amounted to 2,200 acres. Including Searby, which he held in right of his wife, his property extended over 3,600 acres of north Lincolnshire. His farming business was equally impressive. In 1838 he had given up the Nettleton glebe farm of 400 acres, but this loss was balanced by the acquisition of Ewefield and the conversion to arable of a large part of the Moor, and by 1842 the Holton farms covered 1,900 acres. The 1,100 acres that he kept in hand at Searby brought the land that he farmed to a total of 3,000 acres, exclusive of fields hired temporarily for winter feed or summer pasture.

This was an enormous holding by contemporary standards, and probably the largest in north Lincolnshire at that date. Nearly all of it, moreover, was in his own freehold possession, and apart from Searby was largely unencumbered. His mother's death in 1841 relieved the Holton estate of two annual charges, of £145 and £140, and in the early 1840s he was able to restructure his debts to pay off two outstanding mortgages on the Holton estate.[2] In other words, his business was basically sound, and the agricultural improvements in which he had persevered for so many years would begin to pay handsome dividends when prices picked up in the mid-1840s. He did not by any means wipe off all his debts: they still stood at £16,340

---

[1] DIXON 4/14/1–2; 4/15/1.
[2] DIXON 1/E/11.

as late as 1847. But by then they were entirely manageable, and half of them were within the family.

* * *

In the 1840s Lincolnshire entered the railway age. From the middle of the decade Thomas John could shorten his journey to London by travelling by train via Nottingham. But the commercial prize for north Lincolnshire was to establish a direct connection with the markets of Yorkshire, Nottinghamshire and Lancashire, and this was achieved by the line from Grimsby through Gainsborough and Sheffield that became the Manchester, Sheffield and Lincolnshire Railway. The main line across north Lincolnshire ran through Brigg and Kirton-in-Lindsey, but in November 1848 a branch was opened from Barnetby-le-Wold to Lincoln, running through Holton itself.

Thomas John was quick to see the commercial possibilities of these lines, supporting them from the start in 1844. He became a director of the Company in May 1845, and by 1850 owned over £11,000 worth of shares in it.[3] But this enthusiasm did not prevent him from exacting his pound of flesh when it came to the passage of the line through his own property. He obtained the high price of £950 for the 14 acres taken by it, and made various difficulties with John Fowler, the chief engineer. In February 1847 Fowler wrote in exasperation

> We have altered and re-altered the line to please you, we have altered our levels to please you, we have fixed the station to please you, we give an ornamental bridge to please you, and we give you crossings where you ask for them. Now can we do more?[4]

The station was built where the line crossed the Caistor–Market Rasen road, and opposite Thomas John built Holton's first and only public house, the Hope Tavern.[5] Its tenant, Robert Cook, was also given a small farm of 34 acres, taken out of the Stope Hill land, to provide a supplementary income. Behind the tavern a new brickyard was established, and its tenant George Creasey was provided with a cottage.[6] (As with the Stope Hill cottages earlier, it was built just over the Nettleton boundary, to avoid possible Poor Law charges falling on Holton parish.) Further up the line

---

[3] DIXON 22/4/4, fo. 218; George Dow, *Great Central, Volume One, The Progenitors* (London, 1959), 111ff; Neil Wright, *Lincolnshire Towns and Industry 1700–1914* (Lincoln, 1982), 125ff.
[4] DIXON 3/10/3–5; 22/4/4, fo. 225v.
[5] Originally the Hope and Anchor.
[6] DIXON 22/7/5/6. The old brickyard had been rather too near the Hall.

Thomas John made the best of the altered approach to the Hall. The carriage road was taken over the 'ornamental' bridge, and in 1853 a new lodge was erected where it joined the highway to Caistor. 'I would rather have my pretty lodge than the station,' Mrs Dixon had written when the line became a distinct possibility in May 1845. In the end she, and Holton, got both. The hazards of having a railway running through one's plantations were demonstrated in June 1849, when burning coals from a passing locomotive started a small line-side fire.[7]

Parochially the line was, and remained, a nuisance, dividing the Stope Hill fields from Hall Farm.[8] But its wider effects on the Holton district were beneficial, improving its links with Lincoln, Gainsborough, Brigg and Grimsby, although Caistor, marooned on its hill, was a conspicuous loser. For Thomas John the line linked his farms at Holton and Searby – the two stations (Searby was served by Howsham) were only five miles apart – and beyond the county his access to the markets of the Midlands and the north of England was greatly facilitated. By the 1840s he was already sending more wheat and barley to Leeds and Wakefield than to Brigg and Market Rasen, and more sheep and cattle were being despatched to places outside Lincolnshire: the railways enabled him to 'shop around' and respond to the state of the markets more quickly.[9] His sheep, for instance, packed thirty-two to a truck, could be sent to London, to Manchester (via Gainsborough) or to Rotherham (via Kiveton Park).[10] It was a similar story with his imports. By the mid-1840s he was already obtaining quantities of linseed or rape cake, bones and guano from Hull, but these supplies of cattle-feed and fertiliser could now reach him more expeditiously via New Holland and Barnetby.[11] The railways, however, did not totally extinguish the water-borne trade that had been so important to Thomas John's business earlier in the century. Yorkshire coal continued to come up the Ancholme and the Caistor Canal to Riverhead, and it was

---

[7] DIXON 10/3/7; 9/1/47.
[8] It also limited Holton's potential as a shooting estate (information of Mr Philip Gibbons).
[9] In the late 1830s, and again in the late 1840s, Thomas John reverted to local sales for a time.
[10] DIXON 5/3/8, 13–15. For an account of sacks received and despatched, see 6/6/7/6. The railway made mistakes from time to time, losing sacks or sending goods to Holton-le-Clay rather than Holton-le-Moor.
[11] DIXON 22/4/2–4 *passim*.

being consumed in ever greater quantities as the Holton farms went over to thrashing by steam.[12]

In the summer of 1865 the Holton farms were supporting 1,762 sheep, a large flock but not very different from that of twenty years before. The arable acreage, on the other hand, had steadily increased, and by the early 1860s a high proportion of that acreage was devoted to wheat and barley: oats, once an important crop at Holton, had virtually disappeared. In 1862 1,266 acres in all were under arable at Holton and Stope Hill, with 262 laid down to pasture and 141 accounted for by plantations.[13] By this date Searby, always a more productive farm for its size, was producing almost as much wheat as Holton, and its barley, although a smaller crop, appears to have been superior in quality. Only in the size of its flock did Holton continue to be the significantly larger operation.[14]

The expansion of the arable acreage meant a larger workforce. Already by the mid-1840s the Holton farms had a regular complement of about forty-four men and boys, including fourteen confined men on annual contracts. This by no means met the demand for labour at the busiest times of the year. Women and children from the Holton cottages were expected to help out; casual labourers might come from Caistor, Nettleton or North Kelsey; and gangs from farther afield would arrive for the harvest. Michael Sweeney from County Mayo was paid for harvest work from 1845 to 1848, and on 23 July 1849 wrote to Thomas John, probably from Derbyshire, enquiring if the Holton crop was ready for shearing and subscribing himself 'your Honr's obdt. humble servant Michael Sweeney, Irishman'. Dixon replied through his bailiff to say that if he came in a fortnight's time, 'as you have always conducted yourself respectably we will reserve you a fair proportion of harvest work'.[15]

The growth of Dixon's business also involved investment in buildings and machinery. Money was spent at Stope Hill in the early 1840s, Holton itself in the early 1850s (a large barn and granary at Barkworth's) and Beasthorpe in the late 1850s. But, compared with the earlier years, there was no major visible change in the Holton farms between 1845 and 1865. The two significant developments of the period were the increased

---

[12] DIXON 22/4/3, fos 283, 320. A steam thrashing machine was hired from Hart of Brigg in 1843. Clayton and Shuttleworth of Lincoln supplied two engines in 1851. In the 1830s Brandy Wharf, on the Ancholme, had tended to supersede Riverhead as regarded the corn trade.

[13] DIXON 5/1/39; 5/2/15; 6/2/15.

[14] DIXON 5/2/15; 5/5; 6/6/5.

[15] DIXON 22/4/3, fo. 276; 9/6/4/36.

delegation of authority in day-to-day matters and the gradual evolution of the Holton enterprise into a farming-cum-estate business rather than a purely farming one.

As described in Chapter 6, a bailiff had been employed at Holton since 1823. Joseph Mountain occupied that position from 1838, and by 1846 had taken on responsibilities for Searby as well as Holton, at a salary of £100 a year. The job involved handling a large cash balance and keeping meticulous records, and Mountain alas got out of his depth. He went bankrupt, owing his employer £500, and in 1848 was replaced by Henry Seagrave, a young man from a farming family at Lissington, near Wragby. He was engaged at £110 a year, rising to £120 if he gave satisfaction, was allowed a horse, and was given accommodation in part of the old manor house near the Hall.[16] Seagrave was a success. He was given a part-time accountant, Anthony Charles of Nettleton, to help him in 1855;[17] his salary was raised to £170 in 1859; and in 1861 Charles was replaced by a full-time accountant-cum-assistant bailiff, William Lord. (Seagrave was to remain with the Holton estate until his death in 1882, and Lord until his death in 1908.) Further down the chain of responsibility the foremen were key figures at Mount Pleasant, Stope Hill and elsewhere. Robert Favill, the Holton foreman, lodged four men and was allowed to keep two cows. Clearly a capable man, he bought a cottage from Thomas John in 1848 for £60, and appointed Seagrave a trustee in his will. Following his death in 1859 George Maddison, one of the small tenants, was made 'overseer' at Hall Farm.

At that date Holton was still being run as a farm, or group of farms, with only 175 acres let to small tenants. But the estate side of the business was growing in importance. The plantations were sufficiently mature to support auction sales of timber at the Hope Tavern, the proceeds growing from £50 around 1850 to £100 by 1860 and around £240 in the late 1860s.[18] The brickyard near the Tavern could also generate income for the estate. Standard agreements were introduced for the small tenants in the mid-1850s, and an estate survey, the first such comprehensive exercise since the tithe survey twenty years before, was carried out in 1858.

The parish was now quite well wooded. The drive to the Hall had its lodge and ornamental bridge; and the Hall itself began to be referred to,

---

[16] DIXON 22/7/5/5.
[17] Anthony Charles was a local man, the son of Charles Charles of Nettleton, farmer and stock dealer. He may, however, have trained under Wilkinson at Barton.
[18] DIXON 6/6/1; 22/7/5/20–3.

at least by Mrs Dixon, as 'Holton Park'. (Later T.G. Dixon changed the name back to The Hall.) A gamekeeper was employed from 1859, and visitors were occasionally invited over to shoot.[19] But that was as far as the gentrification of the parish went. No further alterations were made to the Hall, and a number of the older cottages remained in their picturesque but insanitary state.[20] The decaying little church was restored – in fact virtually rebuilt – in 1852–4, but it was an undistinguished job, by the Nottingham architect G.G. Place, and cost only £600. Thomas John as lessee of the tithes contributed about £160, the remainder of the sum being raised on the parish rates.[21] The schoolroom was enlarged in 1865. Otherwise the newest buildings clustered modestly around the railway station, about a mile across the park from the Hall.

Between the late 1840s and the early 1860s there were very few additions to the estate. The Beasthorpe leasehold was converted to a freehold in 1854–5, at a cost of £5,020, but an attempt to enfranchise the Dean and Chapter land at Searby-cum-Owmby was unsuccessful. The Holton tithes were purchased outright in the early 1860s, and the Holton land tax redeemed at the same time.[22] But Daisy Hill, the last property in Holton outside the estate, stubbornly failed to come on the market. There was no opportunity to extend the estate into South Kelsey, Owersby or Claxby, all in the hands of large owners, but in Nettleton Dixon took advantage of small opportunities to add to his holdings in the north of the parish, buying 26 acres in 1859 and Bleak House Farm with 60 acres in 1860. The 26 acres were let as small-holdings, but Bleak House was taken in hand and provided with new buildings and a house for a foreman.[23]

* * *

Unlike his father and grandfather Thomas John was late in having children. His elder son Richard came of age in 1851, when his father was already sixty-six, and at the same date Thomas John junior was only sixteen. Neither boy was physically robust, and to their father's disappointment neither showed much interest in farming. Richard was educated with a succession of clergymen. In 1853 he was given a small farm at Searby to manage, but

---

[19] Pheasants were not preserved, but there were hare and partridge, and the gamekeeper could also help to keep down the rabbits.
[20] Amelia Dixon drew some of them.
[21] DIXON 18/6/2; 22/4/4, fo. 113; 22/9/4; 3 DIXON 1/3/4.
[22] DIXON 1/E/12; 22/4/4, fos 118, 147, 237; 22/9/6; Collyer Bristow papers, bundle 166.
[23] DIXON 1/C/7; 22/1/1; 3 DIXON 4/15. At Bleak House a stair went up from the kitchen to the confined men's quarters. There was a similar arrangement at Daisy Hill.

it did not answer, and a few years later John Coverdale, Thomas John's London solicitor, was referring as tactfully as he could to Richard's 'somewhat unsettled and improvident habits'. Thomas John junior went for a spell to the Fauconberge School at Beccles, but in 1851–2 he suffered an illness that must have been mental as well as physical, since Dr Charlesworth of the Lincoln Asylum was among those called in. In 1852 he was copying out some notes on agriculture, but seems never to have taken on any regular work in connection with the farms.[24]

Disappointment was followed by tragedy when in 1854–6 the Dixons lost three of their children to what seems to have been consumption. Charlotte Roadley was brought home from Mrs Leake's school at Panton Hall in July 1854, and died that September. Thomas John junior was then taken by his mother to be nursed at Leamington, but died there early in 1855. A few months later Mary Ann was similarly removed to Scarborough, but died there in September 1856. This left only Richard, Ann and Amelia Margaretta (Emily). Of them Ann was the most capable and the only one to enjoy moderately good health. She finished her education in London, and thereafter became the most widely travelled member of the family, visiting France in 1855 and Germany in 1860. Emily suffered from deafness and a cleft palate, and also, like Richard, from periodic instability. Neither she nor Ann were good-looking, and despite their prospects neither seemed likely to marry (although in the end Amelia did).

More trouble followed. Richard, instead of choosing a bride from within his parents' (admittedly limited) circle, announced his engagement to a Miss Lucy Collinson of Beltoft, in the Isle of Axholme. It is not known how they met: possibly they had shared the same doctor.[25] Her father and her uncle, the solicitor John Collinson, were dead, leaving the family in reduced circumstances, and Lucy's own affairs were complicated, although it eventually turned out that she had an aunt who could come forward with some money for her. For some time Thomas John held out against the marriage, and Mrs Dixon declared that Lucy should never live at Searby.[26] At Coverdale's suggestion, however, William Skipworth, who had connections in the Isle, intervened as a peacemaker. A marriage settlement (of which more later) was drawn up in August 1859, but Richard, already ill from the strain, then suffered a major breakdown, and it was not until

---

[24] DIXON 4/15/5; 9/1/50; 10/5; 3 DIXON 5/12/33.
[25] As suggested to me by Miss Joan Gibbons.
[26] John Collinson had acted for Richard Atkinson of Hatfield, Mrs Marmaduke Dixon's father, a connection that must have revived unpleasant memories at Holton (DIXON 3/3/60–1; 2 TGH 1/16/2).

over a year later that the couple were able to marry.[27] Richard and Lucy settled at Beltoft, but Richard kept in touch with the family at Holton, and eventually Lucy was received there too. For a time Richard's health improved, but he had another bout of illness in 1863–4. Amelia added to these periods of family crisis by becoming ill herself, in 1859 and again in 1863–4; while Ann, torn between family duty and a desire to live her own life, spent periods away from Holton. She and her sister did not get on.

* * *

In 1865 Thomas John reached the age of eighty. It was now clear, if it had not been clear for some years already, that Richard would not take over the farming business. It could not, moreover, be assumed that Seagrave would want to continue indefinitely in his current role of bailiff. He was now in his early forties, and may have indicated that he wished to marry and settle down. It was therefore decided to take the major step of dismembering the farming estate in Holton, Nettleton and Thornton and creating four separate farms, based respectively on the Hall, Ewefield, Stope Hill and Beasthorpe. Hall Farm, a large holding of around 670 acres that included much of Mount Pleasant as well as the land around the Hall, would be kept in hand, but the other three farms would be let to tenants. Seagrave would be promoted to tenant of one of them, but would also act as manager of Hall Farm and agent for the estate as a whole.

The scheme took two years to implement, involving as it did much work in conducting valuations, drawing up agreements, adjusting boundaries, reallocating cottages and modifying buildings. The first farm to be let, Stope Hill, also required the erection of a farmhouse in order to attract a suitable tenant. A pleasant bay-windowed residence standing a little apart from the yard and buildings, it is a good example of what a substantial tenant farmer might expect in the way of accommodation in the high Victorian period. The new occupier was William Storr of South Kelsey, whose father William Storr of Louth had been connected with the Caistor tannery in the 1830s.[28] The marshland that remained in hand was let in the same year, and in 1866 tenants were found for both Ewefield and Beasthorpe, though neither was to reside on the holding. Ewefield was taken by William Goodson of Market Rasen.[29] He was landlord of the

---

[27] DIXON 2/2/15; 3/3 *passim*.
[28] DIXON 22/7/2/73; 22/7/5/14.
[29] He had undertaken the funeral of Thomas John junior in 1855, and catered for the Volunteers' visit to Holton in 1862. He bought a property in Middle Rasen, where he was living in 1872.

White Hart there, a man of capital, and well-known to Dixon. He agreed to take Ewefield plus a part of Mount Pleasant, a total holding of over 400 acres whose fertility was reflected in the rent of £578 a year. Beasthorpe was taken by Seagrave himself, at a rent of £490. This was a commercial rent, but against it he could set the generous salary of £200 a year that he would continue to receive as Dixon's agent and bailiff. That same year he married Eliza Marris from Thoresway, and settled at Kingerby Hall. Kingerby was about six miles by road from Holton, and he rented the Hall from the Young family, for whom he also acted.

The year 1866 also saw the start of the process of letting the Nettleton land apart from Stope Hill. Searby, however, was a more difficult matter. There was no problem about the house, but the farm of 1,050 acres would require a tenant of very considerable capital. In 1866 John William Dixon (James Green Dixon's farming son) offered for it, but he was ruled out by his lack of capital and general unreliability. It was finally taken by William Henry Coates in 1870. He was the son of Thomas Coates of Beelsby, a wealthy north Lincolnshire farmer and landowner, and his wife was a Borman of Swallow, thus connecting him with the Skipworths. He was also a Wesleyan Methodist, but that could be overlooked under the circumstances. He agreed on a rent of £2,000 a year, and had to find £5,730 for the tenant right, of which however £2,000 was left on loan from the estate.[30]

* * *

In divesting himself of these farms Dixon was, obviously enough, reducing his direct income from agriculture, but he was also realising some of the value that he had created in the farms over the previous half-century and more. In fact in the late 1860s his income continued to rise, perhaps to over £8,000 a year gross by 1870, while his expenses shrank. What was he to do with his money? Even in the 1840s, when he had been more debtor than creditor, he had been able to lend comparatively small sums of money to assist friends, neighbours and family members. In 1839 he had lent Thomas Brooks £3,000 to help him purchase the Hundon estate near Caistor. During the 1840s his debtors included his brother James, Henry Green Skipworth of Rothwell and his own bailiff Joseph Mountain. Between 1842 and 1846 he advanced £321 to enable Anthony Bower, son

---

[30] DIXON 22/4/4, fo. 295; 3 DIXON 2/3/4; 3 DIXON 4/3/21. Henry Coates of Grimsby, uncle of W.H. Coates, had supplied Searby with bones in the late 1830s (DIXON 4/14/3; DIXON 15/1: Kirkby pedigrees, vol. 25).

of the Caistor tanner, to go to St John's College Cambridge, a sum that was repaid in 1848.[31] John William Dixon was lent £700 in 1856, and in 1860 even an appeal from the errant George Borman Skipworth received a positive response.[32]

For major investments, however, Dixon turned first to railway shares. His stake in the M., S. and L.R., already mentioned, peaked at £13,000 in 1860, but he also put money into the Great Northern Railway, the Northumberland Dock and, more locally, the Grimsby and Caistor gas companies. From 1863 he favoured the steadier if lower returns from Consols and mortgages, types of investment perhaps more suitable for a family trust. He became mortgagee of the Byron estate at North Killingholme for £10,000 in 1866, and of the Nelson estate at Wyham-cum-Cadeby for £7,000 in 1870, both properties, incidentally, representing in part farming profits made originally on the Brocklesby estate. By 1870 he had almost £50,000 out at interest in the form of mortgages, bonds and shares.[33]

Why had not more of his money gone into land? He was not interested in buying estates outside the county, as Coverdale found when in 1863 he dangled before him the Pennoxstone estate near Ross-on-Wye in Herefordshire.[34] In his immediate neighbourhood, on the other hand, suitable properties came up for sale only infrequently, but he did take advantage of what opportunities there were. In Nettleton he acquired a further 20 acres in the north of the parish in 1865, Wold Farm, a holding of 68 acres in the same area, in 1869, and a public house in the village, the Salutation (formerly the Fighting Cocks) in 1865. More substantial were his purchases in the south of Nettleton parish, adjoining the Holton estate to the east – Oxgangs Farm in 1866, New Farm in 1868 and South Moor Farm (the purchase being completed after his death) in 1871. Including the earlier purchases in 1859–60 these Nettleton holdings amounted to 572 acres, acquired at a total cost of £33,145 10s.[35] In accordance with the policy instituted in 1865 all these properties were let to tenants rather than taken in hand. The largest tenant, Thomas Taylor at New Farm, was inherited from the Eardley (previously Eardley-Smith) estate, which had been broken up following the death of Sir Culling Eardley in 1863. In 1869 Taylor also took Oxgangs, increasing his holding to about 320 acres. During these

[31] Saunders, *Caistor Grammar School*, 20. Bower returned to Caistor as Master of the Grammar School 1853–84.
[32] DIXON 22/4/4, fos 25, 204; 4/15/2; 9/5/10/22.
[33] DIXON 4/15–16.
[34] DIXON 3/4/27–8.
[35] DIXON 1/C *passim*; 22/1.

years there was little scope for extending the Searby estate, but 39 acres in Owmby were acquired from John Ferraby (now of Wootton) in 1866, 17½ acres in North Kelsey at the Nelthorpe sale in 1868, and another 39 acres in Owmby from William Wright of Wold Newton in 1871.[36]

* * *

In his final years Thomas John gave much thought to the question of how all this property should descend to his widow and children. He began making his legal arrangements several years before he dismantled his farming business. To take first the position of his wife, the settlement on Richard's marriage in 1859 confirmed that Searby would be hers for her life after her husband's death: only after her own death would it pass to Richard. During her widowhood it would be charged with £400 a year as an income for Richard, but to compensate her for this deduction a separate settlement of the Beasthorpe farm was executed, under which it would pass to Ann on Thomas John's death but subject to a charge of £400 a year for her mother.[37]

As for the Holton estate, the will that Thomas John had drawn up in 1837 (described in Chapter 6) had to be revisited following the death of Thomas John junior in 1855. Richard was now the only surviving son, and it was decided that he should inherit Holton as well as (eventually) Searby. Accordingly the core Holton estate, that is, all the land belonging to the estate in Holton itself, plus the Stope Hill land in Nettleton, was to be settled on Richard. If he had no progeny it was to pass to each of his sisters in turn. This left a residuary real estate, consisting of all the land not already settled (that is, everything except Searby, Beasthorpe and the core Holton estate), which would be charged with £30,000 in favour of the three girls. Thomas John was thus able to provide handsomely for them without encumbering the core Holton estate or adding to the already heavy charges on Searby.[38]

This will had to be replaced the following year when Mary Ann died. This time the document was a more conventional settlement, and betrays the hand of John Coverdale. It was a distinct possibility that none of the Dixons' children would have children of his or her own, and Coverdale must have pointed out that in that eventuality some male heirs must be put

---

[36] DIXON 22/4/4, fos 144, 147; 3 DIXON 4/13/23.
[37] DIXON 2/1/15; 2/2/15, 17; 3/3/4; Collyer Bristow papers, bundle 165. Mrs Dixon had also been left several thousand pounds by her parents and her husband, adding to her income during her widowhood.
[38] DIXON 3/4/2.

into the succession if Dixon wished the estate to survive into the following generation. The will of 1856 therefore provided that if Richard had no heirs Holton would pass in succession to John William (son of Thomas John's brother James Green), then, failing heirs, to John William's brother Marmaduke, and finally to Thomas Gibbons, the infant second son of the Revd Benjamin Gibbons by his wife Charlotte (née Skipworth) and hence a great-nephew of Thomas John Dixon. Coverdale was familiar with the Gibbons family, having drawn up Benjamin and Charlotte's marriage settlement in 1851. Ann and Amelia, Richard Roadley Dixon's surviving sisters, were to have only £150 a year each out of the Holton and Stope Hill estate, but they would also be entitled to £15,000 each out of the residual estate.[39]

Over the next few years Thomas John continued to ruminate on these matters, and it was not until 1864 that what was to be his final will and testament was drawn up. After Richard's latest illness it was clear that some special arrangements might be necessary, and the 1864 will provided that his trustees would maintain him if he became incapable, paying the rest of the income from the Holton estate to his sisters. Of the male heirs in the will of 1856 Marmaduke, a very successful farmer in New Zealand, had declined the inheritance, and John William had finally been given up as a bad lot. In 1864 Thomas John discarded the Gibbons line and abandoned the idea of male heirs altogether, reverting to a simpler descent within his immediate family. Should there be no grandchildren Ann would now inherit the core estate in fee. The residuary estate, however, was dealt with in a more sophisticated way than before. The land other than Holton, Stope hill and Beasthorpe was to be made into a second *settled* estate, going in the first instance to Ann and then, should she have no offspring, to Amelia. The personal estate (in mortgages and other investments) was to raise £16,000 to ensure that Amelia was properly looked after in her siblings' lifetimes, with the residue invested in land and held for the benefit of Ann. Over the next few years codicils allotted the new purchases of land in the southern part of Nettleton to the core estate, along with the Owmby and North Kelsey acquisitions, but the other Nettleton properties to the residuary estate.

In 1864 Thomas John was still responsible for a very large farming enterprise, and the will of that year was to some extent that of a farmer

[39] DIXON 3/4/3; Collyer Bristow papers, bundle 256. Benjamin Gibbons had no family connection with Lincolnshire. He had become friendly with Charlotte's brother Marmaduke Parkinson Skipworth at Oxford.

rather than that of a squire. The trustees were not clergy or minor gentry but highly respected north Lincolnshire farmers, chosen for their practical wisdom and experience. They were John Iles of Binbrook Hill, William Skipworth of South Kelsey, Thomas Martinson Richardson of Hibaldstow and Thomas Johnson Borman of Swallow.[40] The will directed that the farm account books should go with the estate as heirlooms. A few years earlier Dixon had also wanted to include items of farm machinery in this category, but Coverdale had probably raised an eyebrow at the idea of mangel-wurzel pulpers as heirlooms.[41]

* * *

Servant-keeping was a reliable indicator of status in Victorian England, and it is significant that, just as Holton Hall was very little altered after 1840, neither was the size or smartness of its household. In 1842 the regular staff consisted of a 'houseman' (a superior footman, but below the dignity of a butler), a cook, a housemaid and a kitchen maid. This complement was expanded in later years by the addition of another housemaid, but there was no attempt to emulate the grandeur of Moortown House, which had a butler in the 1850s. While the children at Holton were growing up there were nurse maids and later governesses, but these were temporary expedients. Outdoors the staff did increase a little: by the late 1860s the coachman had been joined by a gardener and a gardener's boy as well as the gamekeeper already mentioned.[42]

What is harder to detect is the degree of formality with which the house was run, and the extent to which the roles of the servants were clearly differentiated. Mrs Dixon continued to be in charge of the household on a daily basis, although in later years she had help with the accounts. In 1845 she complained of a departing cook that she had been 'so fond of the mop and pail that it was a great nuisance'.[43] But in the late 1850s she was going

---

[40] DIXON 2/1/9. Of the four trustees three, Skipworth, Borman (through the Skipworths) and Iles (through the Parkinsons) were family connections. T.M. Richardson had married a niece of George Marris, Dixon's Caistor solicitor. (He was later replaced by Thomas Marris of Ulceby Grange, George's nephew.) All this suggests the hand in Dixon's will of Marris rather than Coverdale.

[41] DIXON 3/4/19.

[42] DIXON 4/18/1–5, 15–17; 10/12/1–5, 20–1; 22/4/4, fo. 200; 22/5/36; 3 DIXON 5/12/1; Jessica Gerard, *Country House Life: Family and Servants 1815–1914* (Oxford, 1994). A lady's maid did join the household in 1857 (*ibid.*, 153), but this was probably for the girls rather than for their mother, since the post was discontinued in 1864. The houseman was called footman in the 1871 Census return.

[43] DIXON 10/3/7.

to a registry office for her cook and housemaids, hiring them on a monthly rather than a yearly basis. They represented a foreign element at Holton, forbidden to 'go about in the village' and inclined to mock the Holtoners on the staff. But the household could not have run smoothly without the help of these village people. Jane Maddison, wife of the coachman David Maddison and formerly Mrs Hewitt, was a key worker at the Hall, doing a range of jobs from ironing and helping to nurse the children to acting as housekeeper in Mrs Dixon's absence. The coachman might be called on to do the marketing, and one of the small tenants (Thomas Hewitt around 1840, George Maddison in the early 1850s) might be involved in accounting for farm produce supplied to the house. (Rabbit and mutton, unsurprisingly, featured regularly on the menu.) Total household expenditure, only about £400 a year around 1830, rose to £1,500 by the late 1850s, but declined in the 1860s with fewer members of the family at home.[44]

In one way, therefore, contacts between Hall and village were maintained, but in other respects the links were weakening. The Dixons, of course, remained responsible for the church, in conjunction with the vicar of Caistor, and for the Sunday school, but as they got older their personal appearances grew less frequent. Thomas John delegated more responsibility to his bailiff, and Mrs Dixon's visits to 'the village' probably became less regular.[45] The agricultural workforce that had worked so closely with William Dixon and the younger Thomas John was losing its older members, and was being more heavily supplemented by casual labour from outside the parish. The changes of the late 1860s meant a further reorganisation of labour, and they also meant the creation of a small group of large tenant farmers, for whom rent dinners were instituted in the schoolroom. But the new tenants, apart from Storr at Stope Hill, were not resident on the Holton estate, and it would take a number of years to build up any *esprit de corps* in the tenant body. The cattle plague of 1866 and a decline in cereal prices did not help things to get off to a good start.

The social vacuum in the parish was partially filled by two nonconformist congregations, a Wesleyan Methodist one served from Market Rasen and a Congregational one in connection with Caistor.[46] But these

---

[44] DIXON 22/5/1. These figures include the children's allowances and holiday expenditure. Mrs Dixon was still compiling the weekly market bills in 1881, when she was over eighty (DIXON 10/12/10).

[45] See Mrs Dixon's journals, DIXON 10/12/1–5. The school treat was maintained as a summer jollity for the village.

[46] R.W. Ambler (ed.), *Lincolnshire Returns of the Census of Religious Worship 1851* (Lincoln Record Society 72 (1979)), 217.

were very small groups, meeting in private houses or cottages, and may not have lasted long. They operated below the radar, as far as the Hall was concerned, and there is not the slightest record of them in the Dixon papers. The village was too small to support much formal life of its own: it is significant that the Foresters who were allowed to hold their summer events in the park came from Owersby, and that the singers who came to the Hall at Christmas were from Claxby and Nettleton. In the railway era Holton did have a public house, but it lay next to the station, at a little distance from the village, and provided the focus for a small sub-community in the parish.

By the 1860s Thomas John had become one of the leading landowners in Nettleton, but he had less influence over its parish affairs than he had had a quarter of a century before. The same was true of Thornton-le-Moor, which for some years had had a resident gentleman farmer, William Parker. Even at Searby the family presence was much reduced. Following the Egginton marriage Mrs Roadley divided her time between her daughters at Holton and North Ferriby (on the Yorkshire bank of the Humber), and Searby Manor was let from 1842. In 1845 the parish of Searby-cum-Owmby, like Thornton and Nettleton before it, acquired a resident parson. The Revd T.J.M. Townsend was an active parish priest, but he found it difficult to counter the Methodist influence that had become established during the incumbencies of his absentee predecessors.

Even during the heyday of the Holton and Searby farms they became less closely integrated with the local economy of the Caistor market area. The number of small transactions with neighbouring farmers declined, and by the mid-century more valuable business was being done with merchants outside the county than with those of Caistor, Brigg and Market Rasen combined. Caistor, as already mentioned, suffered from its lack of railway communication, and during the 1850s its population actually declined. Its resident middle class remained small,[47] and was perhaps less prominent and influential in its neighbourhood than it had been a generation before. James Green Dixon, although financially far from flourishing, continued to play his part in its parochial affairs; and in 1858 a testimonial to him was started. Two years later this led to the opening of a National day school in the town, named after him. But he was not popular in all quarters. Thomas

---

[47] Only nineteen individuals were responsible for nearly half the rateable value of the parish (Rex C. Russell, *Aspects of the History of Caistor 1790–1860, with special attention to the 1850s*, Nettleton, 1992).

John did not cease his connection with Caistor, but it is noteworthy that the Holton family did not subscribe to the testimonial.

* * *

Thomas John's business and social position nevertheless still owed much to family connection. And although of somewhat austere character himself he was a good family man and fond of his children.[48] It must therefore have been very grievous to him and his wife when they lost four of their seven children, and when, of the survivors, one married unsatisfactorily and the other two seemed likely to remain spinsters. The marriage alliances that had strengthened and extended the family network in previous generations were not now taking place. Families either expand or contract, and for the Dixons the period 1840–70 was one of contraction. Old Mrs Dixon died in 1841, her death marking the effective end of the Parkinson connection. Mrs Roadley died in 1858, followed by George Skipworth in 1859. Skipworth had become more gentleman than farmer, but Thomas John had been closely involved with him through business as well as family ties, and his death left an undoubted gap.[49] It was not filled by the next generation: Mrs Skipworth moved to Derbyshire and her unsatisfactory son George Borman took over at Moortown.

These family contractions narrowed the social possibilities of a neighbourhood that had never been crowded at the Dixons' level of society. Ann and Amelia were accomplished girls in their own ways – Ann played the piano and harp and Amelia was artistic – but neither was likely to attract the more eligible bachelors of the Caistor and Market Rasen district. There were quite long connections with the Boucheretts at Willingham House and the Tennyson d'Eyncourts at Bayons Manor, but there was no great intimacy with those families. As for the really grand, a call by Lady Yarborough in November 1870 was a rare honour.[50] From the 1830s onwards, however, contacts with the local professional classes increased. Solicitors

---

[48] Gerard, *Country House Life*, 78–9.

[49] Letters from George Skipworth to his daughter Charlotte (Gibbons), in private possession. Skipworth's contacts ranged more widely outside the county than Dixon's. Of his four daughters (who had portions of £15,000 each), three married clergymen and the fourth a (somewhat eccentric) Lincolnshire landowner, J.L. Fytche. In his last years George Skipworth had a steward at South Kelsey, and employed William Seagrave, Henry's brother, to look after his other properties. The trustees under his will were the Revd J.P. Parkinson of Ravendale, the Revd Lewis Parkin of South Kelsey and a London lawyer (DIXON 2/4/4). George Marris was put out that the trust business, which was to be considerable over the years, went to the London solicitors, Collyer Bristow.

[50] DIXON 10/12/5.

and doctors were often at Holton, and not only on business. And then there was that Victorian phenomenon the gentleman-clergyman, comfortably resident in his parsonage and active in his parish and neighbourhood. Such were Turner of Nettleton and later Maclean of Caistor (from 1844), Hensley of Cabourn, Andrews of Claxby (married to a Skipworth) and Stockdale of Kingerby.[51] The only drawback of these congenial parsons was that they were often on the scrounge on behalf of good causes, of which the most expensive tended to be church restorations. One of Thomas John's last acts was to contribute £30 to the restoration of the church at Thornton-le-Moor.[52]

The Dixons also benefited from the widening of social horizons brought about by the railways. Lincoln was now more accessible for shopping, for meetings, or for calling on Fanny and Martha Dixon, daughters of the Revd Richard, at Deloraine Court near the cathedral.[53] Contact was also made with two minor gentry families near Lincoln, the Bromheads and the Melvilles, leading to invitations to shoot at Holton. Beyond the county Thomas John became familiar with London in a way undreamt of by his forebears, although business trips became less frequent after the 1840s. In the opposite direction trips to Yorkshire became *more* frequent at the same period, the Egginton marriage making Mr and Mrs Dixon honorary Yorkshire people. They attended the union hunt ball at York in 1843, and returned there for the Royal Show in 1848.[54] But it was the watering places that exerted the stronger pull, offering both medical treatment (Thomas John suffered from erysipelas) and congenial company. In June 1843 they found 'a very pleasant company, small but intelligent' at the Crown Hotel, Harrogate.[55] Later they went there mainly in the autumn season, although sometimes they returned from Harrogate feeling more poorly than before, the effect perhaps of the contagious hypochondria of the place. Among seaside resorts Cleethorpes was deserted in the late 1840s in favour of Bridlington, and in the 1850s houses were taken at Filey and Scarborough. Their cousin Dr Kelk practised at the latter resort. Visits to the Yorkshire resorts increased in frequency when the children were ill, with much to-ing and fro-ing especially in the mid–1850s.

---

[51] For Andrews, see Rex C. Russell and Elizabeth Holmes, *Two Hundred Years of Claxby Parish History* (Claxby, 2002), 17–19.
[52] DIXON 22/4/4.
[53] Thomas John left them £100 each in his will (DIXON 2/1/9).
[54] DIXON 9/1/41, 46.
[55] DIXON 9/1/41.

Thomas John never became a magistrate, but in the 1840s he played a moderately active part in the public life of north Lincolnshire, appearing on agricultural, political and protectionist platforms as well as at railway meetings. He usually attended the Assizes at Lincoln, and in 1862 he received the transient and rather expensive honour of being made High Sheriff of the county. This necessitated the acquisition of a coat of arms, and the Heralds' College obliged by supplying the arms of a defunct Cheshire family of Dixon with which he could have had no possible connection, and charging him £76 10s for the privilege. Frederic Burton, the Under-sheriff, wrote in December 1861 to describe the arms: 'The Chief at the top is argent and there are upon it 5 somethings but what I can't make out ….'[56] Mrs Dixon thought her husband looked very dignified when he met the judges at Lincoln. Later in the year he put on two entertainments at Holton. In June 1862 the diocesan architectural society came: one hundred and fifty ladies and gentlemen sat down to luncheon in a tent on the lawn, the band played, and afterwards, recorded Amelia, 'our village labourers partook of the things from the table'. In October it was the turn of the Market Rasen rifle volunteers, but this time the event was not so successful: the tent blew down.[57]

For a younger man the shrievalty might have opened social doors, but Thomas John was by now too old to change his habits. He did however enlarge his subscription list. In 1850 he was paying out only about £20, half of it to bodies such as the Royal and North Lincolnshire agricultural societies, the Caistor ploughing meeting and the local society to restore agricultural protection. The only major charity regularly supported was the Deaf and Dumb Institution at Doncaster, for which he had been a collector in the 1840s. By 1857, however, his annual total of subscriptions had risen to over £30, and it had begun to include county institutions based in Lincoln such as the hospital, the asylum and the Penitent Females' Home.

* * *

The late 1860s were a lonely period at Holton, although the medical crisis of 1863–4 was not repeated. Richard's health, indeed, seemed to improve. Although living at Beltoft he took a certain amount of interest in Searby, and paid visits to both Searby and Holton for the shooting. He sailed a yacht

[56] DIXON 9/6/4/58–9; 3 DIXON 5/11/22–5.
[57] DIXON 10/12; 10/14/1. Thomas John's restoration of Holton church did not evince a great enthusiasm for ecclesiology, but he had for some years been a subscriber to the diocesan architectural society, perhaps at the suggestion of the Revd Richard Garvey, one of Thomas John junior's tutors.

on the Humber and the Ancholme, and even began to play a small part in public affairs. He sat on the Market Rasen Bench from 1869, although he declined the captaincy of the rifle volunteers. Ann was frequently from home. Now approaching her fortieth year, she had an allowance and could look after herself. Amelia also spent several years away from Holton, in the care of her friend Miss Robertson, a former governess at the Hall.

Thomas John's health was good for a man in his early eighties, but it began to fail in the winter of 1870–1. On July 1871, when the Foresters came to Holton, he showed himself at the drawing room window of the Hall, but eight days later he died in the arms of his Caistor physician Dr Jameson. Mrs Dixon, Ann and David Maddison were at the bedside, but not Richard, who left for Filey, presumably in a state of acute distress, shortly afterwards. There he died on 10 August – 'heartbreaking intelligence', recorded his mother, 'and a second overwhelming sorrow to us all'.[58] It must have reminded her painfully of the deaths of her own father and brother, though not in quite such quick succession, nearly half a century before.

[58] DIXON 10/12/5.

8

THE LADIES OF HOLTON, 1871–1906

Thomas John Dixon died a very wealthy man. His total landed estate, settled in various ways, amounted (excluding Searby) to 2,788 acres. His personal property totalled £67,000, out of which he left legacies that included £6,000 to his widow, £10,000 to Ann and £3,000 to Richard's widow. After all deductions there remained a sum of £44,000, out of which £16,000 was to be set aside for Amelia Margaretta. The rest was to be invested for Ann, who as a result of all her father's provisions could expect an income not far short of £2,000 a year during her brother's life. This would have made her a woman of considerable fortune, but Richard's death altered her position again. Her accession to the Holton estate as life-tenant more than doubled her income in the early 1870s, to around £5,000 a year gross, and turned her into a major landed proprietor, with all the responsibilities that entailed. Her mother, too, would be very well off, with the income from the Searby estate and her own capital adding up to around £1,700 a year net of charges.[1] For the time being she was to continue to live at Holton with Ann, contributing £200 a year to the household expenses.

It had been the intention that Thomas John's personalty would be converted into landed property to enhance the Holton estate, but in fact very few opportunities arose to make purchases. In 1875, however, Daisy Hill, the last property in Holton outside the Dixon estate, finally became available following the death of its owner Sir Richard Frederick. The Dixon trustees acquired the 160-acre holding for £8,678.[2] In the same year Owmby Mount, John Ferraby's house near Searby, was bought as a dower house for Mrs Dixon, who under her husband's will was supposed to move out of Holton within three years of his death. Otherwise the money remained as he had left it, in mortgages, railway shares and the like. Byron's mortgage

---

[1] In 1874 the Dean and Chapter leasehold in Owmby and Searby was finally enfranchised, adding another £9,000 to the charges on the Searby estate (DIXON 3/5; 3 DIXON 5/10/14).
[2] DIXON 1/A/7; 22/7/17/2–3; and see also Map 2. Robert Appleton, the Holton shepherd, was sent in disguise to bid for the farm at the auction in Brigg. Needless to say, nobody was fooled.

was called in to finance the Daisy Hill purchase, but the Nelson mortgage was increased to £11,000 in 1877.[3]

After an absence of several years Amelia returned to Holton in the summer of 1875. Over the next few months Dr Jameson, now a widower, was frequently in attendance, and by the end of the year he had proposed for her. Mrs Dixon, with what misgivings are unrecorded, agreed to the marriage, and Ann bought the couple Holly House in Caistor, a house similar in date, design and appearance to Holton Hall. Following the marriage in October 1876 there were unpleasant revelations, probably to do with Jameson's debts, but after this bumpy start the family settled down to the new regime. Mrs Dixon decided not to move to Owmby Mount, but to divide her time between Holton and Caistor, very much as her mother had lived alternately with her own two daughters.[4] The Owmby house was renovated and let to Septimus Skipworth, a distant cousin and sanitary inspector for the Caistor union. He paid the rent erratically but regularly complained about the drains.

Mrs Dixon died, well into her eighties, in 1885. Her will, drawn up ten years previously, used her power of appointment under the 1859 settlement to leave the Searby estate to Ann. Amelia, however, was to have half the income from it, and was additionally left the sum of £2,000. Should she become incapable of managing her affairs through 'mental infirmity' the trustees were to maintain her and pay the remainder of her income to Ann.[5] Ann was now owner or life-tenant of nearly 4,500 acres, and, it might be supposed, even better off than she had been in the early 1870s. In truth several years of agricultural depression had greatly reduced her income. The Holton estate and home farm were barely breaking even, and Searby, although in a somewhat better state, was clearing no more than about £300 a year by the mid-1880s. It was fortunate indeed that she was a spinster of no extravagant habits, and that her income was not solely derived from agricultural rents.[6]

When she came to make her own will in 1889 she was free, under her father's will, to make a new settlement of the Holton estate, to cater for the fact that both she and Amelia were childless. As its ultimate heirs she rejected, as her father had, the progeny of her uncle James Green, but reverted to the Gibbons relations first considered over thirty years before.

[3] DIXON 22/4/14, 16.
[4] DIXON 10/12/7–8.
[5] DIXON 2/1/12. Amelia's possible incapacity was covered by a codicil of 1878.
[6] For Holton and Searby estate and farm accounts and papers, see DIXON 5/1/46–72; 6/1/5, 12; 6/4; 6/5/1; 6/6/8; 6/7; 22/4; 22/7.

On Amelia's death, therefore, Holton would pass to Cousin Charlotte, Mrs Benjamin Gibbons, and then to her second son Thomas George (Tom). Successive Gibbons heirs were to take the name of Dixon in lieu of Gibbons when they inherited. The property settled was to include the 'second' or residuary estate and the trust fund as well as the 'core' Holton estate. It had already been arranged, however, that the Beasthorpe farm would pass to Amelia absolutely, while the Searby estate would on Amelia's death pass under a separate deed of appointment to Charlotte Egginton's side of the family.[7]

Ann survived her mother by eight years – lonely years spent partly away from Holton – before dying in 1893. She was then succeeded at both Holton and Searby by Amelia Margaretta, who now called herself Mrs Jameson Dixon. Dr Jameson had died in 1890, but she had continued to live at Holly House, and moved to Holton only in 1895. By then farming had improved a little, and Amelia enjoyed an income somewhat greater than had Ann in her last years. When she herself died in 1906 Holton passed to Tom Gibbons, his elderly mother having declined the inheritance; Beasthorpe was sold; and Searby passed to a Mrs Martin, Charlotte Egginton's eldest daughter, who sold it the following year.

* * *

By 1906 the Holton and Searby estates had survived a deep and prolonged depression in the farming industry. They had done so, moreover, in the care of three ladies, none of whom was an outstanding businesswoman, none of whom enjoyed robust health, and none of whom had a close uncle, brother or nephew on whom she could lean. They did, however, have a coterie of advisers willing to assume some responsibility for the welfare of the family and the estate. Ultimate legal responsibility, of course, was vested in the two sets of trustees. For Holton Thomas John had selected four experienced farmers, but of these William Skipworth had predeceased him and Borman, the most active of the remaining trustees, retired owing to ill health in 1878. For the next ten years the leading trustee was the Revd John Hodgson Iles, son of John Iles of Binbrook Hill and sometime archdeacon of South Staffordshire. Although only a distant relative he took a conscientious interest not only in Holton but also in Searby, for which he

---

[7] The probate of Ann's will was preserved by the solicitors to the Holton trustees, but the Searby deed of appointment would have been with Crust, Todd and Mills, and has not been traced. It may have been handed over when the estate was sold.

had been made a trustee in 1875.[8] Mrs Dixon's other Searby trustee was also a clergyman, the Revd William Waldo Cooper of West Rasen, son of that Mr Cooper who forty years before had been so doubtful of Thomas John's suitability for the Lindsey Bench.[9] When Ann drew up her will in 1889 Archdeacon Iles had recently died, but she appointed to the trust his son John Cyril Iles, Cooper and the Revd John Francis Quirk, of Grasby and later Great Coates near Grimsby. Quirk was not a relation and not even a native of Lincolnshire, but he nevertheless took on the role of leading trustee and adviser to the family.

It was, however, the lawyers who kept the estates running legally and financially, and who indeed helped to choose the trustees. For fifteen years after Thomas John's death Marris and Smith remained responsible for the Holton trust, but in 1886 it was decided to transfer that business to the Lincoln solicitor H.K. Hebb. Hebb was an energetic and successful man (he was also Town Clerk of Lincoln), and it was he who drew up Ann's will. But it was also necessary that Mrs Dixon and her daughters should have their personal legal advisers, and that role fell to Messrs Crust, Todd and Mills of Beverley. This Yorkshire firm had been the solicitors for the Searby estate since the 1820s, and it was natural that Mrs Dixon should stay with it during her widowhood. Thomas Crust drew up her will, and later James Mills advised Ann and Amelia in connection with Holton as well as Searby. There were difficulties between the two sets of lawyers from time to time, over questions such as whether the trust or the life tenant should be responsible for a particular item of estate expenditure. Mills (who was Town Clerk of Beverley) was not always willing to concede a point to Hebb, and letters flew between Beverley and Lincoln, adding to the bills of charges as they did so.

Crust, and later Mills, paid periodic visits to Holton, but it was on their bailiff or agent that the ladies relied for the day-to-day management of their properties. Henry Seagrave continued for a decade after Thomas John's death in his multiple role of bailiff, agent and tenant farmer: he represented a reassuring element of continuity in the family's affairs. But he never wholly grew into the role of agent, and when Storr gave up Stope Hill in 1880 he chose to add it to the home farm rather than re-let it (always supposing, of course, that he could have found a tenant). He died, suddenly, in January 1882, and was succeeded by his nephew and protégé

---

[8] Mrs Dixon's will of 1875 left £100 each to her trustees. Ann left £5,000 to Mrs Iles, the Archdeacon's widow.
[9] See above, Chapter 6.

George Marris. Ann, who had found Seagrave increasingly resistant to change, was probably not sorry. But Marris was no more a professional agent than his uncle: his real interests were in farming and horse-breeding. He moved into Stope Hill, and became not only tenant of that farm but also of Mount Pleasant and (on Mrs Seagrave's retirement) Beasthorpe.

So matters went on in Annie's lifetime, but thereafter relations between Marris and Mills rapidly deteriorated, with Mills having to point out that if Marris disregarded the terms of his own tenancy he could hardly expect the other tenants to respect theirs.[10] In 1895 he was deprived of the agency and Beasthorpe, and he gave up the other two farms in 1897. Mills took the opportunity to introduce a proper agent, in the form of W.H. Todd of Hull, a cousin of his partner William Todd of Beverley. W.H. Todd introduced new agreements, took a tougher line on arrears, and found new tenants for the farms. The home farm was reduced to a mere 78 acres in 1897, and the long series of Holton farming day books came to an end. Now, finally, Holton had become a conventional tenanted landed estate.

Ann had dealt generously with her farming tenants, letting their arrears mount up and excusing them some of their debts to the estate rather than turn them out of their holdings. A permanent reduction of ten per cent was granted in 1882. At the end of the 1880s the estate finances showed some signs of improvement, but in 1892 large sums in arrears were written off, and ten per cent was returned at the Lady Day audit in 1893. When Amelia took over she was warned by Mills that strict economy would be necessary 'for some time to come'.[11] The other way in which the estate tried to keep its tenants was to spend money on permanent improvements. Under-draining was carried out, and farm buildings repaired or extended, but the most conspicuous expenditure was on farm houses. Stope Hill, always a difficult farm to let, had already been given a house in Thomas John's time. Yewfield (as Ewefield was now spelt) received a good house in 1887. And Beasthorpe, with its heavy clay soil, got a house in 1899, although this time, since it was Amelia's own property, she had to meet the expense out of her private account. 'You will notice', wrote Todd to Mrs Jameson Dixon about the plans in December 1899, 'that there is not much provision for [a] flower garden as farmers of 300 acres do not usually desire it, but there is provision for [a] useful kitchen garden and berry bushes, etc, with

---

[10] DIXON 6/7/23.
[11] 3 DIXON 5/10/16.

just a nice little grass plot in the front of the house and a drive to the front door and a cartway to the coalhouse and back door.'[12]

And what – apart of course from Seagrave and Marris – of the tenants themselves? The first generation at Holton were local people, but, with the possible exception of Goodson, not men of very large capital, and when corn prices fell and harvests failed they were soon in difficulties. The only large tenant to weather the storm was Coates at Searby, and he did so by alternately appealing to Ann's sympathy and threatening to quit. His rent was reduced by half over the years, and at that level he probably continued to make ends meet. Searby Manor suited him and his family well enough, and he stayed there even after succeeding to his father's estates at Beelsby and Hatcliffe in 1887.[13] At the other end of the scale the Cook family also managed to keep going during the depression. Tom Cook added a local coal merchant's business, based in the station yard, to his public house and small farm, and by 1906 was the only tenant farming on the Holton estate whose family went back to Thomas John's time.[14] A very different type of farmer from either Coates or Cook was Jessop Hargrave, the son of a butcher and stock dealer from the Market Rasen district, who took on Yewfield in 1887 and made it pay by fattening large numbers of cattle on the farm during the summer. His son John became the largest tenant on the Holton estate in the 1890s, occupying Mount Pleasant as well as Yewfield. He farmed extensively rather than intensively, economising on labour and letting some of his least productive fields go out of cultivation.[15]

Despite the limitations imposed on Ann by her gender and her legal position she was by no means a cipher as a rural landlady. She had inherited some practical commonsense from her father, and an interest in philanthropy from her grandfather. Unlike either of them, however, she had a strong concern for the physical well-being of her cottage tenants and their families, and her first project on inheriting Holton was to do something about the poor state of the parish's housing stock. She went with Borman to look at the Yarborough cottages at Claxby and the Angerstein

---

[12] DIXON 10/6/2.

[13] In January 1897, during one of the estate's periodic tussles with Coates, Searby Manor Farm was widely advertised not only in the Lincolnshire and Yorkshire papers but in the *North British Agriculturist* (DIXON 22/4/13).

[14] Lincs. Archives, LAMB 11. He held 167 acres by 1905. His coal business, at first very local, later included Caistor and Market Rasen.

[15] Hargrave papers, in private possession. Miss Joan Gibbons of Holton, the author of *The Flora of Lincolnshire,* remembered that when her family first came to Holton and John Hargrave was still farming at Yewfield, one of his fields was full of excellent wild flowers.

ones at Owersby, and appears to have decided that the latter provided a better model.[16] The old house near the Hall, no longer needed as a bailiff's house, was given a slate roof in place of its thatch and turned into two three-bedroomed cottages, and west of the main street several new cottages were erected between 1872 and 1883.[17] They were attractive as well as comfortable, and enhanced the appearance of the village. Less visible but equally important was its water supply. Ann had been distressed when babies had died from drinking brackish water during the hot, dry summer of 1868, and she determined to lay on piped water from the Wolds at Nettleton Hill. She is said to have been opposed by Seagrave but supported by her trustees. The scheme, carried out in 1874, cost around £1,000, and a ceremony was held to mark its completion.[18]

Amelia never saw eye to eye with her sister, so it is not surprising that when she succeeded she did not continue with the cottage-building. (She probably shared her mother's appreciation of picturesque if bug-ridden thatch.) She did extend the piped water to Daisy Hill, which had still been off the estate in 1874, but otherwise she contented herself with a few small improvements near the Hall – an enlargement of the lodge, a new fence around the front of the house and swans for the pond. In one respect, however, she showed herself a true granddaughter of William Dixon, leaving money in her will for the erection of almshouses at Holton. They continue in use today as a form of sheltered accommodation for older people connected with the estate.

* * *

One might have expected the late nineteenth century to be a golden age for Holton as an estate village. The ladies at the Hall had all the accepted notions about supporting and patronising the church, school and other village institutions.[19] And they had the means to carry out their charitable intentions. But in fact Holton was not a thriving, perhaps not even a

---

[16] DIXON 10/12/6.
[17] Later Tom Dixon gave them names associated with the history of Holton (Drewry, Willoughby, Stovin, etc.).
[18] Miss Joan Gibbons was told that Mrs Dixon drew the first glass. A service of thanksgiving and a dinner for the villagers in the long granary followed the ceremony. The engineer for the work was George Bower of Huntingdon, son of the Revd Anthony Bower of Caistor. Jane Brand, a former Holton schoolmistress, wrote to T.G. Dixon in 1921 that Ann's 'aim and object always was to fit her people for their walk of life and perhaps few know how much she did for her fellow men' (DIXON 11/5/2/83).
[19] See Pamela Horn, *Ladies of the Manor: Wives and Daughters in Country House Society 1830–1918* (Stroud, 1991).

very contented, community during these years. Ann was often away – in 1877 she got as far as Italy – and when at home she did not do much entertaining. She and her mother dispensed with a houseman in 1873, and maintained a modest female staff. Amelia enjoyed her role as lady of the manor: she claimed it gave her a local rank 'next to the Queen'. But she lived quietly, almost reclusively, at Holton, protected by a small band of faithful retainers. She kept only a housekeeper, a cook and a maid indoors, and on one occasion excused herself from receiving the vicar of Caistor on the grounds of 'having no housemaid to wait at the door'.[20] Ann had continued to rely on the Maddisons, but Amelia installed her own servants from Caistor. Mrs Mumby was the housekeeper, her brother Charles Welch the coachman, her husband the manager of the little home farm, and her brother-in-law William Pike the gamekeeper. The Mumbys' daughter Minnie was a favourite of Mrs Jameson Dixon's, and was to be remembered very generously in her will.

The family continued to take a close interest in both the church and the school, but in both cases the influence of the Hall was circumscribed. Canon Westbrooke, vicar of Caistor from 1886, took a more active part in Holton affairs than his predecessors, but Amelia in particular did not like his advanced High Church views. Presumably most of the curates who appeared to take services at Holton were High Church too.[21] The Sunday school became a day school in 1873, with financial support from Ann, and by 1884 it had over thirty pupils. In that year she enlarged the schoolroom, previously the Sunday school. Archdeacon Iles, and later Canon Quirk, were encouraged to visit it. But as a day school it now had a full-time mistress, whose ideas and methods did not always give satisfaction at the Hall.

The population of the village shrank during the depression, from 195 in 1871 to 167 in 1891, despite the new cottages. Some of the cottagers now worked for the farmers, rather than directly for the Dixons; and farmers such as the Hargraves were not likely to contribute a great deal to the village community. By 1900, too, there were more inhabitants who did not work on the land at all – the schoolmistress, the railway workers, the policeman and one or two men connected with the Nettleton ironstone mine. The area round the station, moreover, continued its development as a little sub-community within the parish. In 1881 the schoolmistress,

---

[20] DIXON 10/14/3. When Tom Gibbons (later Dixon), staying at Claxby for his cousin Andrews's funeral, tried to call he was told that Amelia was 'engaged for six weeks' (information of Miss Joan Gibbons).
[21] For Westbrooke, see David Saunders, *Caistor's Vicar and Church 1886–1916* (printed Heighington, 2003).

who had fallen out of favour at the Hall, was put up by her friends the Cooks at the Hope Tavern when she left her job, and in order to avoid possible embarrassment her successor was met off the train at Moortown rather than at Holton.[22] Some years later Tom Cook caused direct offence by erecting a memorial to his sister in Holton churchyard in the form of a cross. Mrs Jameson Dixon thought this was too High Church and also 'unsuitable for a poor person', and sent Todd and Lord to inspect it, but it was of course outside her jurisdiction.[23]

The links that connected the parish and its leading family to the wider world of the district and the county had also weakened. Ann could not occupy the public positions that had come almost automatically to her more recent forebears. She did, however, attain a certain prominence as a supporter of local good causes. In 1874 she gave £2,000 towards a wing of the County Hospital in memory of her father, and later contributed a further £400 towards its completion.[24] In 1878, remembering her brother's interest in boats, she funded the purchase of a lifeboat for the Lindsey coast. A site was found at Donna Nook, near the family property at Skidbrook, and a tablet was erected there 'in dear remembrance of Richard Roadley Dixon – to God alone be the glory'. The station turned out to be inconveniently placed, and was closed in 1931.[25]

* * *

By 1906 the great days of Thomas John Dixon's farming enterprise at Holton may still have been remembered by a few local people (including old William Lord, with his now white whiskers and increasingly red nose), but not by many. And in that year Thomas John's Holton estate passed out of the hands of his direct descendants. It was not the outcome that he had worked and planned for. Yet the remarkable thing was not what had been lost but what had been preserved. The estate would not have survived had not Ann and Amelia been able to subsidise it from their own resources, and they would not have had those resources had their father not left a personal fortune in addition to his landed property. The trustees had worried from

---

[22] DIXON 10/6/1/8.
[23] DIXON 10/14/3.
[24] DIXON 11/6/3.
[25] 3 DIXON 5/12/17; *Lincolnshire Standard*, 17 June 1988; Jeff Morris, *The Story of the Mablethorpe and North Lincolnshire Lifeboats* (R.N.L.I., 1989). The boat for Donna Nook cost £363 and its carriage £139.

time to time that they had not carried out his instructions in putting all his money into land, but their failure to do so was the saving of the estate.[26]

The fate that could have overtaken Holton can be gauged from two examples in its immediate neighbourhood. It may be recalled that James Green Dixon, Thomas John's youngest brother, had inherited Gravel Hill Farm in Thornton-le-Moor, and had set himself up as a landowning farmer on the Wolds at Rothwell, the other side of Caistor. In 1849, at the age of sixty, he had handed over Gravel Hill to his sons John William and Thomas Parkinson Dixon; and following the death of his wife in 1872 he sold a half-share in Rothwell to his children. But he had let the Rothwell farm run down, and its mortgage run up. His wife's fortune went to his children, not to him, and when he died in 1879 his estate was scarcely sufficient to pay his debts. John William had continued to farm at Gravel Hill (Thomas Parkinson having been committed to an asylum), and in 1873 he also took on Rothwell. But he had never been a competent farmer, and the bad times that set in at the end of the 1870s, plus the growing burden of debt on the Rothwell farm, before long rendered his position hopeless. On his father's death he had moved into the White House at Caistor. Becoming increasingly eccentric, he embraced the British Israelites, and his wife was said to have hidden in the cellar to escape his monologues on the Lost Tribes.[27] Rothwell was sold up in 1885, leaving little even for the wretched mortgagee; Thornton was abandoned in 1890; and at that point John William and his immediate family left the county.[28]

John William thoroughly alienated his cousin Ann Dixon, but she was kind to his youngest brother James Green Dixon junior, who had been unwelcome at the White House after his father's death. A simple soul, and unable to support himself, he remained in Caistor in a small cottage, and was left an allowance in Ann's will. He died in 1918, aged eighty-six, when the Revd T.G. Dixon, acting as his executor, salvaged his papers and those of his father and added them to the family archive at Holton.[29]

It may also be recalled that George Skipworth, of Moortown House, South Kelsey, had died in 1859. He had held only a life interest in the South Kelsey estate, which had been settled under his father's will and which carried a £50,000 mortgage. But he had built up a personal estate of 832

---

[26] Annie indemnified them in a deed of 1889 (Sills and Betteridge papers, bundle 1801).
[27] Information from the Misses Gibbons.
[28] DIXON 12/4/7/40, 42; 12/9/1/8–11; 20/1/13; 2 TGH 1/25/5. Three of the sons went to New Zealand.
[29] DIXON 12. Gravel Hill was reacquired for the Holton estate in the late twentieth century.

acres, worth £1,800 a year by 1859, and he left in addition a considerable fortune in ready money. These circumstances, apart from the mortgage on the main estate, bore some resemblance to the position at Holton. But there the similarity ended. South Kelsey was entailed on George's surviving son George Borman Skipworth (1820–1890), whom George disliked and in whose business abilities he had, rightly, no faith. He therefore left all he could away from his son and in trust for his widow and daughters. George Borman inherited a large estate but no money to reduce the encumbrances on it or provide a cushion in hard times. In 1869 it had a rental of about £6,000 but fixed charges of around £4,000. By 1882 the rental had fallen to £4,000 but the mortgage had risen to £108,000. As at Rothwell this was an unsustainable position. The South Kelsey estate took time to go under, but following George Borman's death in 1890 it passed into the hands of the mortgagees, and was re-sold in 1902.[30]

In both cases a combination of personal and financial failure had precipitated the decline of a once prosperous family. But there were two further parallels. G.B. Skipworth, like J.W. Dixon, felt that he had been harshly treated by his family and to some extent disinherited. Dixon's mind turned to the Lost Tribes, Skipworth's to the cause of the Tichborne Claimant, one of whose most ardent supporters he became. But the Skipworths always did things on a larger scale. While John William was merely spotted on a soapbox at Hyde Park Corner, George Borman went to prison for three months for contempt of court.[31] The other connection is that in both cases one thread in the story led back to Holton. Tom Dixon was, through his mother, one of the ultimate heirs of the trust set up under George Skipworth's will: he inherited some property in Caistor, and was also able to obtain some relevant papers from the firm of Collyer Bristow, successors to Eyre and Coverdale, who had been administering the trust for the previous half-century.[32] Thus papers for the Skipworths as well as for James Green Dixon and his family now form sections of the Holton archive.

---

[30] Lincs. Archives, LPC 3/3/1.
[31] Douglas Woodruff, *The Tichborne Claimant: A Victorian Mystery* (London, 1957), 242–3.
[32] See DIXON 13. Lincs. Archives BRA 1498 is a deposit from Collyer Bristow through the British Records Association.

9

FARMING AND LANDOWNING

> When I canters my 'erse along the ramper I 'ears
> 'proputty, proputty, proputty'.[1]

This chapter aims to provide a broader context for the history of the Dixon family, by exploring the extent to which the larger farmers of north Lincolnshire became owners as well as occupiers of land in the eighteenth and nineteenth centuries. Why did some buy land on a large scale, while others did not, and how did some farming families, having acquired estates, manage to keep them for more than one or two generations?

Human nature is seldom straightforward, and behind each purchase of an individual field or small estate there might be a mixture of motives. These motives, moreover, might have elements of sentiment as well as hard-headed practicality. Land, as discussed in the next chapter, had a social value in parts of the countryside. And farmers could take into account the future of their families as well as their immediate business requirements. The present chapter, however, concentrates on financial and prudential considerations.

Even in practical terms a farmer could have a number of reasons for buying land. He might want to take it in hand, or use it to set up a son in business. He might buy a field or two for winter feed or summer grazing. He might want to expand a holding already in his possession or occupation, or prevent land from falling into other and possibly undesirable hands. Alternatively he might want to use his purchase not for farming but for the purposes of investment – as a security of whose value he was a better judge than many. The story of the Dixons has shown how there could be more than one reason for a particular purchase, and how land acquired for one purpose might be retained for a different one.

As every farmer knew, the land market, like any other market, was subject to the laws of supply and demand. As far as supply was concerned, the two significant factors were price and availability. In eighteenth-century Lincolnshire there was a large difference in price between rich lowland

---

[1] This is Alfred Tennyson's 'new-style' northern farmer speaking (quoted by Christopher Ricks in *The Poems of Tennyson* (London, 1969), 1189).

and unimproved upland.[2] Even where land was comparatively cheap to buy it might require substantial investment before it could be made to yield a profit. Nevertheless bargains could still be picked up in the early nineteenth century, as shown by the South Kelsey and Althorpe speculations.[3] By 1850, however, such ventures had become a thing of the past. The rise in the price of farmland was assisted by its increasing 'amenity' value, and working farmers could find themselves in competition with rich outsiders wanting to move into the countryside. An estate of 1,000 acres could now cost £35,000. If let at 28 shillings an acre it might yield a return of four per cent, but that was before any outgoings.

Price was of course linked to availability. In certain parts of the county it was rare for land to come on to the market in any quantity, either in complete estates or split into smaller lots. If it did it was seldom at the right time or in the right place as far as farming purchasers were concerned. If a chance did arise the competition might simply be too strong. In 1842 Lord Willoughby de Eresby decided to sell Withcall, near Louth, an entire parish of over 2,000 acres occupied mainly in one very large farm. The tenant, Richard Dawson, thought he had 'as much claim ... as his Lordship' to Withcall, but he could not raise the asking price of some £80,000, and it went instead to Colonel Tomline of Riby Grove.[4] (Riby itself, incidentally, was not to be sold until the 1930s.[5]) Withcall had been an outlier of the very extensive Willoughby estates in Lindsey, and it did happen occasionally that such outliers would be sold off, sometimes in lots. Thomas John Dixon, as recorded in Chapter 7, was able to take advantage of sales of land from the Eardley and Nelthorpe estates in the late 1860s; and earlier James Green Dixon had been able to create a large farm from three adjacent holdings on the periphery of the Brocklesby estate. But such opportunities were rare.[6]

If the supply of land was becoming increasingly limited, it can also be suggested that demand for it from farmers was becoming weaker. In the days of William Dixon senior there had been relatively few opportunities for the wealthy rural investor, apart from land and mortgages on land. By

---

[2] Holderness, 'English land market', 557–76.
[3] See Chapters 4 and 6 above.
[4] Lincs. Archives, Ancaster papers, 3 ANC 7/23/43; Olney, *Rural Society*, 56.
[5] Riby Grove was gutted and demolished in 1936–7 (DIXON 20/3/6–7).
[6] Earlier examples included the sale of the Wootton and Kirmington estates of the Winn family in the late eighteenth century, enabling the Hudson, Nicholson and Brooks families to acquire land on the edges of the Brocklesby estate, and the sale of the Twistleton estate in Barnetby-le-Wold in 1820.

1850, when Thomas John was building up his portfolio of railway and other shares, it was safer – as long as one avoided the high-risk companies – to invest one's money in such securities than it had been a generation or two before. It was also more flexible than tying up one's capital in land, especially since the practice of high farming was demanding more and more working capital.[7] There might be times in a farmer's life, as William Dixon found in 1798, when ready money was 'more preferable than land'.

Another reason for owning one's farm rather than renting it, and one would have thought a telling one, was security. The early nineteenth-century Lincolnshire farmer was more likely to be an annual tenant-at-will than a leaseholder; and it was not pleasant to have a notice to quit drop unexpectedly through one's letter-box. But paradoxically the growing amount of tenant capital embarked in agricultural holdings strengthened rather than weakened the position of the tenant. Landlords became keener to attract and *retain* wealthy tenants, and by 1870 evictions were rare on the large Lincolnshire estates. Moreover, if a tenant did leave his holding for any reason he could claim compensation for some of the money invested in his farm under the well-established Lincolnshire Custom.[8]

Confidence between landlord and tenant was a particular feature of the Brocklesby estate, where farms were frequently passed down in families, and where it was not unknown for a tenant to appoint trustees in his will to continue the tenancy during the minority of an 'heir'.[9] During the 1870s one Yarborough tenant, William Frankish, provided for his sons not by buying farms for them but by taking additional holdings on the estate, increasing his total tenancy from 450 to 1,986 acres and 'embarking' £18,000 in the family enterprise.[10] Another leading Brocklesby family, the Goultons, farmed at Croxby for over a century without ever becoming landowners.[11] Members of the Brooks family, one branch of which held the other large farm at Croxby, did buy or inherit land at various dates in the late eighteenth and nineteenth centuries, but always in holdings of less – usually much less – than 1,000 acres; and it was their tenure of farms

---

[7] E.P. Squarey, 'Farm capital', *Journal of the Royal Agricultural Society*, 2nd Series XIV (1878), 425.
[8] Olney, *Lincolnshire Politics*, 40–3.
[9] See, for instance, the wills of William Richardson of Limber 1830 (Lincs. Archives, LCC Wills 1830/193) and Philip Skipworth of Laceby 1836 (DIXON 2/4/3).
[10] *Parliamentary Papers*, 1881 xvii, 652.
[11] Lincs. Archives, BRA 1216.

on the Brocklesby estate that characterised their family history more than their ownership of land in its own right.[12]

* * *

All this suggests that we shall not discover a large number of Lincolnshire farmers buying substantial quantities of land in their native county between, say, 1725 and 1875. But quantifying this is difficult.[13] In an article published in 1974 B.A. Holderness concluded that the eighteenth-century Lincolnshire land market provided farmers with opportunities for making purchases, but that they, the farmers, did so more often for investment than for farming purposes, and that the land so acquired tended to stay in the family for no more than two generations.[14] Unfortunately, however, without sufficient documentary evidence for individual farmers and their families it is often impossible to say with certainty how much land was acquired and when, how much of the money came from farming profits, and how long the property remained in the family. The Dixon collection, in this as in other respects, is exceptional.

Holderness's study, moreover, does not take us far into the nineteenth century. But there are two sources that make the early 1870s a useful point of reference. One is the enumerators' returns to the 1871 Census, which for working farmers give the size of their holdings; and the other is the 1873 Return of Owners of Land. These two sources are not precisely contemporary, and neither is perfectly accurate: sometimes the census enumerators fail to include the farming acreage, and the 1873 Return, based on the efforts of parish overseers, is of variably quality. But White's 1872 *Directory of Lincolnshire* is an invaluable supplementary source, and can in some cases clarify whether land owned in 1873 was in owner-occupation or let to tenants.

---

[12] DIXON 15/1, vol. 25, p. 10; information kindly supplied by Mrs Susan Ellis of Thornton-le-Moor and Mrs Joan Mostyn-Lewis (later Russell) of Claxby. It should be noted, however, that members of both the Richardson and the Brooks *families* between them owned a total of over 1,000 acres in 1873.

[13] Land tax duplicates 1759–1832 survive among the Lindsey Quarter Sessions records, but they are incomplete and do not give acreages or occupations.

[14] Holderness, 'English land market', 558. He found an 'impressive' number of farming families who rose into the 'minor gentry' in the eighteenth century (570), and calculates that about twenty such families owned over 1,000 acres by 1800 and a further six over 2,000 (571). But he does not define 'minor gentry', and it is not clear from his account exactly who these families were. Understandably his local knowledge is better for south Lindsey, the area chosen for his doctoral research, than for the area which features more prominently in this discussion.

Table 1 takes the Census as its starting point, and analyses the evidence not for the whole division of Lindsey but for three unions, Brigg, Louth and Caistor, that together covered much of the northern Wolds. The Census yields a total sample of 590 farmers holding 200 acres and upwards in 1871. Of these as many as 357, or approximately three-fifths of the total, owned a negligible or non-existent amount of land (less than one acre) in 1873. Roughly another fifth (113 individuals) owned between one and 99 acres, and only the remaining fifth (120 individuals) owned 100 acres or more. (See the bottom line of the table.)

Table 1
Farmers Owning Land in Three Lindsey Unions, 1871–3

Occupying	Owning		
	Less than 1 acre	1 – 99 acres	100 acres and over
200 – 299 acres	120 (62%)	50 (25%)	24 (13%)
300 – 399 acres	78 (68%)	21 (19%)	15 (13%)
400 – 599 acres	98 (62%)	21 (13%)	38 (24.5%)
600 acres and over	61 (49%)	21 (17%)	43 (34%)
Total	357 (60.5%)	113 (19%)	120 (20.5%)

Sources: Census enumerators' returns 1871 (TNA, RG 10/3398–3438); Return of Owners of Land (*Parliamentary Papers*, 1874 lxxii); White's *Directory of Lincolnshire* (1872).

Within the sample as a whole it might be expected that the larger the occupier the more land he would be likely to own. This was the case, but only to a certain extent. Only among farmers of 600 acres and upwards did the proportion owning any land at all rise to just over half (64 out of 125). (And only among the occupiers of 1,000 acres or more, not shown in the table, did it rise again, to 64%.) In terms of the size of estates owned, it was only among the farmers occupying 400 acres and above that the proportion owning more than 100 acres rose to over one-fifth (24.5%).[15] The corresponding figure for those occupying 600 acres or more was 34%,

---

[15] One might have expected 300 rather than 400 acres to be the more significant figure. According to Davidoff and Hall (*Family Fortunes*, 24) farmers of over 300 acres were more likely to exhibit middle-class characteristics, including the choice of land or securities rather than houses or buildings as favoured types of investment. In Lincolnshire too it has been noted that of 94 south Lindsey members of the North Lincolnshire Agricultural Society in 1860 only 15 farmed less than 300 acres (James Obelkevich, *Religion and Rural Society: South Lindsey 1825–1875* (Oxford, 1976), 51).

and for those occupying 1,000 acres or more 60%. That is a considerable increase, it is true, but it is clear that even among the agricultural plutocrats landowning on any scale was far from universal.

Of the 590 farmers in the sample, it has been possible to identify 79 who were tenants on the Brocklesby estate. Despite the fact that these included some of the wealthiest farmers in the district, only fifteen owned any land, a proportion (37%) slightly *lower* than that for the sample as a whole (39.5%).[16] It is true that at the very top of the farming scale the Brocklesby farmers featured somewhat more prominently: of the eight who occupied over 1,000 acres, six owned more than one acre. But only one, William Wright of Wold Newton, *owned* over 1,000 acres in 1873: he had recently bought Wold Newton from the estate.[17]

At the 1,000-acre level these figures for farmers become of no statistical significance, but it becomes useful to look more closely at individual farmers such as William Wright and their family backgrounds. By changing our focus to the Return of 1873, we can attempt to isolate a cohort of landowners at that date who were either farmers themselves or whose families had risen through agricultural profits to become substantial landowners in earlier generations.

The Return produces a total of 131 individuals or families owning 1,000 or more acres in north Lincolnshire (the administrative division of Lindsey) in 1873.[18] Of these, however, 46 represented old county families settled in the county (mainly in north Lincolnshire) by 1775. A further 53 represented families that had moved into the division after 1775, but had owed their rise into the 1,000-acre bracket to money made, wholly or mainly, from sources other than farming. This leaves a small total of 22 owners who, or whose families, had become substantial landowners through money made wholly or mainly from farming in the division after 1775. These owners are given in Table 2, where they are subdivided into those with estates acquired after 1820 (group A), and those with estates already totalling 1,000 acres at that date (group B).

---

[16] Lincs. Archives, YARB 5/2/4/1: rental for the Brocklesby estate 1862. The size of the estate alone, some 55,000 acres, ensured that its tenants are well represented in the larger sample.

[17] Wright had become the owner of the Wold Newton estate by 1873, but the Return fails to register this. Thomas John Dixon had himself recorded as still farming 1,000 acres in the 1871 census, although this was no longer the case.

[18] The Return does not distinguish the three divisions of the county, but the Lindsey proprietors have been identified with the aid of county directories and personal knowledge. The table of course excludes those who both acquired and lost their estates in the century before 1873.

Table 2
Farming or Ex-Farming Families Owning Over 1,000 Acres in Lindsey, 1873

	Acreage in 1873	Farming in 1873	JP in family
A. Acquiring 1,000 acres after 1820			
BENNARD of Owmby	1,019	+	x
BRIGGS of Oxcombe	1,021	+	x
VESSEY of Welton-le-Wold (i)	1,039	+	x
HILL of Winceby	1,241	+	x
CHATTERTON of Stenigot	1,262	+	x
GRANT of Hareby	1,308	+	x
WHITLAM of Biscathorpe	1,322	x	x
ALLENBY of Cadwell	1,431	x	+
SOWERBY of Beelsby	1,526	+	x
NELSON of Wyham	1,688	+	x
COATES of Beelsby (ii)	1,821	+	x
HUDSON of Kirmington	1,988	+	x
WRIGHT of Wold Newton	2,000 (iii)	+	x
WELLS-COLE of Fenton	2,231	x	x
SLATER of North Carlton	2,650	+	x
B. Already owning 1,000 acres in 1820			
MAW of Cleatham	1,124	x	x
OTTER of Ranby	1,225	x	x
ROADLEY of Searby (iv)	1,484	x	+
WRIGHT of Brattleby	2,200	x	+
DIXON of Holton-le-Moor	2,275 (v)	x	+
PARKINSON of Ravendale	2,901	x	+
SKIPWORTH of South Kelsey	5,542	x	+

Notes to Table 2
(i) In addition to 1,039 acres for J.H. Vessey of Welton-le-Wold, the Return gives 289 acres for his father Samuel 'Vissey' of Halton Holegate. The Vesseys were land agents as well as farmers.
(ii) Thomas Coates and his relative and neighbour John Sowerby jointly owned the Beelsby and Hatcliffe estates. In addition Coates had land at Waltham, Holton-le-Clay and Langton-by-Horncastle. The farmers who *occupied* over 1,000 acres in 1871 also included W.H. Coates of Searby and T. Sowerby of Withcall, but neither of these owned any land. The

Notes to Table 2 continued

Coates and Sowerby families had benefited from joint business interests in Hull. Neither was an old north Lincolnshire farming family.

(iii) Approximate figure.

(iv) Entered under 'Mrs Dixon'.

(v) This acreage (the core settled estate) is correctly given to Ann. The Return however appears to omit Beasthorpe Farm and all but 43 acres (attributed to T.J. Dixon's executors) of the 'second' or residuary estate. The Return was often puzzled by land held in trust or by partners or co-heirs, reminding one of the difficulties encountered by present-day compilers of 'rich lists'.

It is not surprising that as many as 15 out of these 22 landowning families had acquired their estates, or at least the properties that took them over the 2,000-acre mark, after 1820 (see group A). The great majority of these estates in fact represent single purchases, whole lordships acquired with the intention of being farmed by their new owners.[19] Most were still being run as owner-occupied farms in the early 1870s. (Two owners had retired from farming, and in the third case, that of Biscathorpe, there had been a failure of male heirs.) In at least four cases – Oxcombe, Winceby, North Carlton and Wold Newton – the purchaser was the sitting tenant.

Most of the estates in group A were, however, under 2,000 acres. In considering the handful of families in this group that owned around 2,000 acres or more, it becomes clear that they had exceptional resources. Either they had already entered the land market before making the purchase that took them up to that size, or their fortunes were based partly on inherited land or money as well as farming profits. Thus Samuel Slater was already a substantial owner, having inherited land in the Winterton area jointly with his brother J.B. Slater, when he bought North Carlton from the fifth Baron Monson in 1838.[20] William Wright of Wold Newton, a cousin of the Brattleby Wrights, inherited land at Owmby, near Searby, that he sold to help finance the purchase of Wold Newton from the Yarborough estate in 1870.[21] William Hudson (c.1793–1881) had a more scattered estate, and one that owed more to inheritance: he had received land in Kirmington and Barnetby-le-Wold from his father, and had also benefited from the wills of his two elder brothers.[22] William Wells-Cole of Fenton (d. 1867), who had

---

[19] The exception was the Bennard family, which built up its holding over two generations.

[20] TLE 40; MON 25/13/14. Slater paid £77,000 for the 1,800-acre estate, including the advowson of the living and a handsome Elizabethan manor house. A leading sheep breeder, he disposed of his prize flock to help raise the purchase price (Perkins, *Sheep Farming*, 49).

[21] The Owmby land was bought by John Bennard (see note 19 above).

[22] Lincs. Archives, Sills and Betteridge deposit, SB/Foster: Foster family deeds.

farmed under Lord Yarborough at Newstead, near Brigg, inherited land from no fewer than three farming families, the Wellses of Dunstall (near Gainsborough), the Coles of Fenton and the Clarkes of Brumby.[23]

This leaves a select group of seven families (group B) that had already entered the 1,000-acre bracket by 1820. Of them five did so during the period of the French Wars, 1793–1815. In three cases out of the five an element of speculation may have been involved. By far the largest acquisition was that of the South Kelsey estate, already referred to in Chapter 4.[24] A partner at one time of Philip Skipworth's in the purchase was Francis Otter, who later bought the Ranby estate but spent his last years farming at Stainton-le-Vale.[25] George Maw was farming at Bigby, near Brigg, when in 1799 he entered into a partnership with his brother Matthew, the Brigg corn merchant, to buy the Cleatham estate. This suggests a speculative aspect to the transaction, although when Matthew died he left his share to George's son, another Matthew.[26]

The other two families that made major purchases during the French Wars had already become landowners on a significant scale before 1790, and their war-time acquisitions had more to do with farming and estate improvement than with short-term profits. The Roadleys, as recounted in Chapter 3, were settled at Messingham long before Richard Roadley bought the Searby estate. The Wright family also made money from farming and bought land during the pre-war years, although it was William Wright (1750–1797) who made the important purchase. He was farming at Somerby (next to Searby) when in 1795 he purchased the large and as yet unenclosed Brattleby estate, a few miles north of Lincoln.[27]

Just two families in group B remain to be considered, the Parkinsons and the Dixons. Like the Wrights they could trace their north Lincolnshire origins back to the late sixteenth century, and like the Wrights again they emerged as substantial landowners in the eighteenth. Unlike the Wrights,

---

[23] Information kindly supplied by Cdr Peter Wells-Cole. Estate, family and legal papers were deposited in Lincolnshire Archives in 1975 as WELLS-COLE (*Archivists' Report* 25 (1973–5), 39).

[24] Philip Skipworth made money from sheep-breeding as well as land deals (Perkins, *Sheep Farming*, 49). Holderness ('Land market', 572) states that he also profited as a contractor for enclosure works.

[25] TLE 40/3; MISC DON 264/16; DIXON 1/F/1/9; TDE F/1–2; LCC Wills 1814/185. The Otters originated from the Nottinghamshire side of the Trent, and had links with the Raynes family.

[26] North East Lincolnshire Archives, Maw of Cleatham papers, 379/A 28, 34. Later both the Oxcombe and the Stenigot estates were purchased jointly by two brothers, with one subsequently buying the other out.

[27] Lincs. Archives, MISC DEP 118; DIXON 15/1, vol. 25, p. 116.

however, they made no major purchases during the period 1793–1815. The Parkinsons were not actively engaged in agriculture by that time, while for the Dixons it was more a case of a lack of suitable purchasing opportunities. As we saw in Chapter 3, the Ravendale Parkinsons were the senior line of a family that had owned land in both the Scunthorpe and the Healing neighbourhoods in the early eighteenth century: it was the junior, Healing, branch that became so closely connected with the Dixon family. The Ravendale branch was connected with the Wrights, Ursula Parkinson marrying William Wright (later of Brattleby) in 1782.[28] The purchase *and retention* of land was a traditional characteristic of all these interconnected families. The local and family connections of all the families in group B is remarkable.

Of the major purchases Ranby lay near Wragby and Brattleby near Lincoln, but in both cases the purchasers had links with the Brigg–Caistor–Market Rasen district. And it was in that district that the other five all lived. (One cannot quite say that all had their principal properties there, because Ravendale was not owned but merely occupied by the Parkinsons at this period.) It can also be noted that Cleatham, Searby, South Kelsey and Holton all lay in the lowland rather than the upland part of the district, although Searby, it is true, included wold land in the parish. Money could be made on the Wolds, but the opportunities for investing it in land were more likely to occur in the lowlands.[29]

The Dixons and Parkinsons were also unusual among the families in group B in accumulating their estates piecemeal, both before and after 1820, rather than rising into the 1,000-acre league through one very large transaction. The Parkinsons, however, ended up with very scattered estates, at least partly due to the fact that no land was available, until a late date, in the immediate vicinity of their seat at Ravendale, whereas the Dixons were able to build up an estate in the very neighbourhood in which they had their farming business. Their most remarkable achievement was to restore the previously divided parish of Holton-le-Moor to single ownership. As already described, they were able to do this by buying farms and cottages as they came on the market, avoiding for the most part the strain on family finances that could occur when a desirable but very large estate came up for sale at an inconvenient period of the family's history.

---

[28] North East Lincs. Archives, Parkinson of Ravendale papers, 361, 542.
[29] On the east side of the Wolds the parishes of Barnoldby-le-Beck, Laceby, Healing, North Killingholme and Immingham presented somewhat similar opportunities.

This again was the consequence of the peculiar tenurial history of Holton, a history, however, that was not entirely unique in the district. The small estate of Kingerby, near Market Rasen, was divided by sales in the mid-seventeenth century, but in the period 1785–98 it was reassembled by the Youngs, a farming family long resident in the neighbouring parish of West Rasen.[30] James Young (1757–1823) built Kingerby Hall (the house much later occupied by Henry Seagrave, the Holton agent), and died in possession of about 935 acres.[31]

But perhaps what most distinguishes the Dixons from the other families in group B is that the head of the family continued in full-time farming during the period 1820–65. Around 1860 the heads of the Maw, Otter, Wright, Parkinson and Skipworth families were living in some comfort in their small country houses, mostly built in the early years of the century, and leaving others to farm their land – although in the case of Maw this was his son rather than an ordinary tenant. Thomas John Dixon, on the other hand, was not only still farming but had greatly increased his operations by taking on Searby in addition to Holton. Around 1860 he occupied 3,000 acres. Only two other north Lincolnshire farmers held over 2,000 acres at that date, Sowerby of Withcall (at Dawson's old farm) and Torr of Riby, but neither of them was also a landed proprietor.

Can the families in Table 2 tell us something not only about how some farmers came to own sizeable estates but also about how they were able to pass them down the generations? Three related factors seem to have been crucial: family size, patterns of inheritance and the existence or otherwise in the family of non-landed wealth. As regards family size, ideally there would be one competent son to inherit in each generation. But in practice there would often be too many children or too few. This problem also affected aristocratic families, but the difference was that they tended to believe in primogeniture, whereas farmers favoured partible inheritance. We have seen how Thomas Dixon divided his properties among his children, and how his eldest son, William junior, divided his own land between his two farming sons. In a similar fashion Robert Parkinson of Ravendale (1690–1740) left Scunthorpe to his eldest son and Healing to his second son. In the cases of both the Parkinsons and the Dixons this did not cripple

---

[30] Lincs. Archives, 4 BM 4/1–2; Richard Olney, 'The Youngs at Kingerby: the making, un-making and re-making of a Lincolnshire estate', in Christopher Sturman, ed., *Lincolnshire People and Places: Essays in Memory of Terence R. Leach* (Lincoln, 1996), 117–20.

[31] The Kingerby estate, even had it exceeded 1,000 acres, could not have been included in our sample, since it had been broken up by 1873.

the senior branch. But it helped with the Parkinsons that the two estates were not intertwined for farming purposes, and with the Dixons that the eldest son had the means and determination to recover from the setback of the parental dispositions.

Sometimes a parent would not divide his property physically, but would saddle it with charges in the interests of family fairness. In the eighteenth century John Grant made a fortune from farming at Oxcombe, in the southern Wolds, as his monument in the parish church proudly proclaims. He left the Oxcombe estate to his five sons as tenants in common, directing that one of them should farm it as tenant to the other four.[32] But by the 1840s it had passed to the Briggs family. At Kingerby James Young left the estate to his eldest son in 1823, but charged it with £20,000 in favour of his two younger sons. Fraternal disputes led to a partition in 1838, leaving the eldest son with the house but only half the land. Following his own death in 1875 his portion was also divided, and sold as a consequence of further legal and family difficulties.[33]

If, in contrast, there was a failure of heirs one might think that the failure of the estate would be even more likely, but in fact such was not always the case. Between 1840 and 1875 the Ravendale and Holton estates descended to daughters, and Brattleby to a cousin, yet all three estates survived into the twentieth century. In the case of the Dixons this was due, as we have seen, to the use of settlements to preserve the inheritance, and the same was no doubt true of the other two. Owners such as Thomas John Dixon were not necessarily converts to aristocratic doctrines and practices, but when circumstances called for it they would go to a good lawyer and tie up their properties in at least a quasi-aristocratic manner.

The third factor, the availability in a family of money that was not tied up in land, is not always obvious from the surviving evidence, but it played a crucial role in the survival of both gentry and farming families. An estate bought with borrowed money in good times would remain viable as long as the good times continued and rents held up. If agriculture entered a period of depression, as it did after 1875 and again at the end of the 1920s, rents would fall but the mortgage payments would remain the same. Other outgoings might even increase. In these circumstances we have seen how vital an additional fund in shares and other assets could be: the Holton estate survived, whereas the South Kelsey estate, heavily mortgaged and with no other source of income, was in the hands of the mortgagees by

---

[32] LCC Wills 1799/i/97. Oxcombe had been purchased in 1785.
[33] Olney, 'The Youngs at Kingerby'.

1900.³⁴ Of the fifteen estates in group A, eight had gone by 1900, and a further four by 1933.³⁵ At the latter date only the Whitlam, Wright of Wold Newton and Hudson estates had survived, almost certainly assisted by the operation of family settlements. But the first two were by then in a parlous state. The third, represented by the Foster and Neale families, owed its survival at least partly to an injection of industrial wealth.³⁶ Of the families in group B, only the Skipworths had become landless by 1900, but Searby and Ranby were sold early in the twentieth century. By 1933 the Dixons were still at Holton, the Maws at Cleatham, the Wrights at Brattleby and the Parkinsons at Ravendale, but only the Dixons had managed to retain more or less intact the estates recorded in the Return of sixty years before: the others had had to reduce their holdings by sale.

It would be helpful to compare the Dixons with other farming and landowning families outside Lincolnshire, but published studies are few, and the survival of archival material on which such studies could be based is equally rare. Two examples must suffice here, one from north Northumberland and one from south Norfolk.³⁷

Matthew Culley (1732–1804) and his younger brother George (1735–1813) were the sons of a County Durham farmer but moved to Northumberland, where, with a third brother, they established a successful farming partnership, most notably as sheep-breeders. They took and improved various leasehold farms, and between 1795 and 1807 invested large sums in the purchase of land in the Wooler district of north Northumberland. In fact they founded two gentry families: in 1873 the descendants of Matthew and George both owned estates of between 2,500 and 3,000 acres in their adopted county. George wrote to his son in 1813 about the transformation

---

34 For the fate of James Green Dixon's properties see above, Chapter 6.
35 George Nelson of Great Limber (1771–1848) was said to have made £88,000 by farming (Olney, *Lincolnshire Politics*, 39 and n.), but there was a £7,000 mortgage (to Thomas John Dixon) on the Wyham and Cadeby estate in 1870, later increased to £11,000. The estate appears to have been sold after the early death of Nelson's great-grandson in 1885 (DIXON 15/1, vol. 25, p. 119).
36 William Hudson's niece and principal heir married the Lincoln iron founder and thrashing machine maker Henry Munk Foster: it was he who purchased the Searby estate in 1907. Industrial wealth also played its part in the reconstruction of the Kingerby estate, by another branch of the Young family, in the late nineteenth and early twentieth century (Olney, 'The Youngs at Kingerby', 119–20).
37 D.J. Rowe, 'The Culleys, Northumberland farmers 1767–1813', *Agricultural History Review* 19 (1971), 174; Anne Orde (ed.), *Matthew and George Culley: Farming Letters* (Surtees Society 210, 2006); Susanna Wade Martins and Tom Williamson (eds), *The Farming Journals of Randall Burroughes (1794–1799)* (Norfolk Record Society 58, 1995).

of the family fortunes: fifty years before he had 'worked harder than any servant we now have, and even drove a coal cart', yet here was his son living in a 'palace' (Fowberry Tower).[38]

The Wymondham district of south Norfolk, with its smaller and more intensively farmed holdings, was a very different environment from that of north Northumberland. Yet at the end of the eighteenth century it also produced a very successful farmer, Randall Burroughes of Burfield. Intelligent and meticulous, he built up his business, like William Dixon, as a mixture of tenant, owner-occupier and rentier landlord, and by 1806 had acquired an estate of just over 1,000 acres in the large parish of Wymondham. I have not discovered what happened to his own estate, but the family of his elder brother, the Burrougheses of Burlingham, became prominent Norfolk gentry during the nineteenth century.[39]

These two examples appear to underline the importance of the period 1793–1815 as one when agricultural fortunes could be made with *comparative* ease. They also attest to the importance of inheritance when it came to the formation and preservation of landed estates. Matthew Culley married a local heiress, and the estate owned by his descendants in 1873 had come to the family by inheritance. Likewise Randall Burroughes appears to have inherited some family money, and he too made a good marriage: Burfield came to him from his wife. His antecedents, moreover, were different from those of the average Lincolnshire farmer. He was the well-educated son of a gentleman with both land and brewery money in the family, and although as hard-working as any practical agriculturist he could be described by that rather treacherous phrase 'gentleman farmer'. Even the Culleys, come to that, were scarcely self-made. Their father, a landowning farmer with good connections in County Durham, had given them an excellent start in life by sending them as pupils to Robert Bakewell of Dishley Grange, Leicestershire, the foremost breeder of his day. The Culleys' correspondence reveals a breadth of outlook that was a world, or at least several counties, away from the north Lincolnshire circles in which the Dixons moved.

* * *

This chapter has established that it was unusual for the larger farmers of north Lincolnshire to purchase significant quantities of land in their own

---

[38] See Burke's *Landed Gentry*, 1937 edn.
[39] *Ibid.*

county.[40] Exceptionally successful farmers might occasionally buy large farms, of over 1,000 acres, to take in hand. In a few fortunate cases they were able to buy as sitting tenants. But to acquire 2,000 or more acres was highly unusual, and could hardly be done in one generation, or without the help of inheritance or money from elsewhere. Once acquired these estates might stay in the same family for a generation or two, but very seldom beyond that. Cyclical downturns in agriculture, a failure of competent heirs or the effects of the practice of partible inheritance all worked against the long-term retention of such estates.

Against this background the achievement of the Dixon family comes into sharper focus. In each generation, from William senior to Thomas John, those mainly responsible for carrying on the family business were at the least frugal and competent, at the most energetic and successful to an exceptional degree. They worked through the bad times and took advantage of the good. They were moreover lucky in that no head of the family died prematurely, or died without children. In 1798 and again in 1824 the family property was divided, and there was unhappiness over the inheritance, but not to the extent of tearing the family apart. In the nineteenth century cross-breeding gave way to in-breeding, and Ann and Amelia, the children of first cousins, had no offspring themselves. But at that stage, as we have seen, the estate could survive through a combination of careful management, the fortunate existence of spare capital, and the operation of a family settlement under which it would pass to male cousins.

In all this the part played by habit and family tradition should not be underestimated. It took over a century to consolidate the Holton estate, and that gave plenty of time for the sentiment of 'keeping the name on the land' to be established and strengthened. Even so the career of Thomas Dixon (1729–1798) is a reminder that not every generation may run true to type. Thomas was not born at Holton, nor did he spend most of his adult life there. Instead he developed an allegiance to the neighbourhood in which he farmed for many years, and specifically to the landlord of whom he was an adviser as well as a leading tenant. Like a number of his neighbours who farmed under the Pelhams of Brocklesby he showed no interest

---

[40] It was rare for wealthy businessmen generally to buy large estates. See W.D. Rubinstein, 'Business-men into landowners: the question revisited', in Negley Harte and Roland Quinault (eds), *Land and Society in Britain, 1700–1914* (Manchester, 1996), 90–118. His statistics, however, refer to very wealthy businessmen and very large estates. F.M.L. Thompson points out (*ibid.*) that one needs to count estates purchased in the generation after the original fortune was made as well as those purchased by the businessmen themselves.

in buying land, and seemingly no hankering after the independence that owning one's own farm might be thought to confer.

This in turn is a reminder of the fact that land, though fundamental to the rural society of north Lincolnshire, did not by itself determine either the social class of those who owned or farmed it or the nature and extent of the local communities to which they belonged. It is to those social, as opposed to economic, aspects of our subject that we return in the final chapter.

10

CLASS AND COMMUNITY

In the century from around 1730 to around 1830 the fortunes of the Dixon family were transformed. In the former year William Dixon senior was a substantial tenant farmer, but owned no landed property except what had recently come to him through his wife. A hundred years later Thomas John Dixon was not only farming on a much more extensive scale but possessed over £50,000 worth of land. One might expect there to be a clear correlation between wealth and status, and that it would be easy enough to show that as the family became steadily more prosperous so it made an even ascent through successive layers of rural society.

Life, however, at least for the Dixons, was not so simple. To start with, their accumulation of wealth was not a steady process. William senior certainly became a wealthy man by contemporary standards, owning 1,750 acres and enjoying an annual income of about £500 by the time of his death. But when his son Thomas died less than twenty years later, he had not added to that estate (unless one counts his wife's Keelby farm), and the expenses of his large family had circumscribed his income, reducing it to a level not a great deal higher than his father's. William junior, as we have seen, built up his own farming business during his father's lifetime, and made purchases to extend the Holton estate; but as head of the family in the early nineteenth century he let his own business decline, concentrating instead on assisting his children and on pursuing his charitable work. When he died in 1824 his *net* income was less than his grandfather's fifty years before. It was left to Thomas John to raise the family fortunes to a much higher level. During his father's lifetime he seized the favourable opportunities of the war-time period, and through his marriage became both a landowner and a farmer on an impressive scale.

In simple terms of wealth, therefore, the family experienced no steady rise to prosperity in the period 1730–1830. In social terms it is hard in some respects to detect any significant 'rise' at all. During these years the Dixons remained working farmers. They began their careers by learning from their fathers or other close relatives as apprentices or junior partners. Then, in their mid-twenties, they set up in business on their own account, usually with parental help. They started on good-sized holdings, or at least holdings sufficient to support the wife and child or children for whom they were now responsible. The figure of 700 acres seems for decades to have been a bench-mark in the family, although a 700-acre mixed farm in the 1790s was of course a much more substantial concern than a grazing holding of a similar size half a century earlier. Once established in this way it was usual for the head of the family to continue working, though not necessarily in the same farm, until his death (Thomas) or until an advanced age (William senior, Thomas John). Even William junior did not finally retire until very near the end of his life, although he had been reducing his holding for some years.

The assumption that son would succeed father in the family business, if not in the same farm, led to other assumptions that had social implications. The choice of schooling for the son or sons intended for the agricultural profession was of particular significance. The local grammar school turned out young men who could write a fair hand, reckon a column of figures and carry themselves creditably in the kind of circles in which they would mix. But – unless given a fuller classical education, as William junior's clergyman brothers were – the products of Brigg or Caistor Grammar School seldom emerged into their local world with more than a local outlook, or with much intellectual curiosity. (William Dixon junior, exceptionally, had this curiosity, but it was that of an autodidact conscious of his educational limitations.) Later in the life of the working farmer his narrowness of outlook would be reinforced by the demands of a profession that left little time for travel or leisure. Thomas John was the exception here. As a young man he did travel, and later he had a bailiff who could mind his farms for him if he took the occasional holiday.

Narrowness of outlook and lack of leisure in their turn restricted the suitability of the working farmer, however experienced and competent, when it came to filling local appointments. And this was important in terms of local standing, since some of these appointments not only recognised status but also served to enhance it. William senior, Thomas and Thomas John all became tax commissioners, a post for which they were qualified by property and by personal reputation.[1] But the duties were not very onerous or time-consuming. A seat on the county Bench was a different matter. It had a much greater *cachet* – a tax commissioner would be styled gentleman in official documents whereas a magistrate would be addressed as esquire – but to be a useful Justice of the Peace, and one

---

[1] William junior was the exception, but he contributed to the public life of the district in other ways, to the point of neglecting his own interests.

moreover acceptable to one's fellow-magistrates, required both more social polish and more leisure than the working farmer could usually command. Thomas, exceptionally, *was* made a J.P., but his usefulness was confined mainly to his own locality. Thomas John, despite his wealth and ability, was not, although it is true that by that period the social bar had been raised.[2]

Young farmers might generally be expected to choose brides from the same level of the farming class as themselves. After all, the wife of a working farmer would not be a lady of leisure, or at least unadulterated leisure, and would find her new life difficult unless she had come from a farming background herself. But here this rather undynamic account of the Dixon family history needs modification. Down to Thomas John's generation the Dixons who farmed did take brides from what may broadly be called the local farming community. Yet in each generation the wife was of slightly higher status than her husband, and more clearly of higher status than her mother-in-law. William senior's wife brought him only a modest property, but she had wider family connections than her husband, and came, unusually, from (just) outside the county. Thomas's wife was no heiress, but she was of very good north Lincolnshire farming stock, and her father was also a clergyman – again a first for the Dixon family. William junior's wife was not only a clergyman's daughter with a portion of £1,500 but a member of an old-established propertied family, some of whose property, indeed, descended to her sons. Mary Ann Roadley was an heiress on a different scale: she was a farmer's daughter, and of course a Dixon on her mother's side, but she brought to Holton not only a modicum of superior sophistication but a fortune of about £30,000.[3]

Turning to the younger sons and daughters, the marriages of Thomas's younger children helped to consolidate the social position of the family. The two younger sons became parsons, and of the daughters two married very respectable north Lincolnshire merchants, the third a young farmer with excellent prospects and the fourth a doctor. William junior, and

---

[2] In the early 1870s it was still most unusual for a working farmer to be considered for the county Bench (Olney, *Rural Society*, 102). Of the landowners in group A of Table 2 (see p. 161), only Samuel Allenby was a magistrate, and he was a retired farmer with gentry connections who kept a genteel establishment at Cadwell Hall. Most of the families in group B had supplied magistrates at various times, but in 1872 Ann Dixon was disqualified as a woman, Francis Otter as a retired solicitor, and Maw and Skipworth as eccentrics.

[3] Searby and Messingham together were valued at £80,000, but Mary Ann's net share had to allow for the charge in favour of Mrs Roadley.

seemingly his wife, had no social pretensions, but their only surviving daughter made the best marriage in the family to that date: her husband was a farmer, certainly, but also the heir to a large estate. William junior's two younger sons became a solicitor and a corn merchant, and married respectively a solicitor's daughter and a corn merchant's daughter.

This social movement was reflected in the houses and households of the family. William senior had lived as a widower in a small house at Holton attended, we believe, by only a housekeeper. Thomas and his wife began their married life in somewhat uncomfortable circumstances at Firsby, but from 1758 lived in a handsome modern house at Riby. Mrs Dixon had the use of a chaise and groom, although her indoor establishment was restricted to two female servants. Their eldest son was the first to occupy the new manor-cum-farmhouse built by Thomas at Holton: no doubt his wife's family would have expected no less, and she like her mother-in-law had a chaise for paying visits. But she, again like her mother-in-law, made do with two maids. It was to this, by then somewhat old-fashioned, house, that Thomas John brought his bride in 1827.[4] But this time changes were made. Mary Ann began married life with three rather than two maids, and soon the house itself, as well as the household, was enlarged and improved to meet the needs of the growing family. As for visible marks of status outside the home, it was not long before the chaise was upgraded to a phaeton (a four-wheeled carriage), and the footman to a coachman.

So far this chapter has discussed the Dixons in terms of their rank and status within their own rural society. But how far, and at what period, can they be located within a wider English middle *class*? As described in Chapter 1, there are difficulties in transposing a predominantly urban concept into a rural setting. William Dixon senior is a particularly problematic case, partly owing to a dearth of evidence, but partly also to his very attenuated family circle and the thinness of his local society. By the end of the eighteenth century, however, one can say with more confidence that the Dixons belonged to a local middling sort. Thomas and his wife at Riby were anchored in a society in which their position was indicated by their comfortable if modest household, their attitudes to family and family property and their connections with a neighbourhood made up of merchants and professional people as well as farmers. Their son William and his wife at Holton were part of a very similar neighbourhood, although their house was family-owned, not rented, and it lay at the centre of a

---

[4] The more up-to-date style of the period was represented by houses such as Searby, Moortown and Kingerby.

growing estate. Even so, one hesitates to describe the family at this period in class terms: as shown in Chapter 5, William himself avoided the word in his analyses of local society.

By 1830, however, we see in Thomas John and his wife a couple not only on a higher plane of prosperity but one that, with its higher standards of domestic comfort, London-acquired furniture and travel beyond the county, was embracing a more national middle-class culture. Nor, of course, did its social journey stop there. In 1830 Thomas John could already boast landed assets double those held by his father. Forty years later those assets had more than doubled again, and he was enjoying an income that could easily support a member of the landed gentry. But is that what he became?

By 1860 the establishment at Holton Hall (or Park) included a gamekeeper as well as a coachman and a gardener, and the social distance between the house and the village had widened. The village and parish looked more like the property of a squire than it had thirty years before. In terms of personal standing Thomas John had never joined the Bench, but he was about to become High Sheriff, and his son was in due course to become a magistrate. When it came to making his will he consulted a London lawyer, and his dispositions followed aristocratic conventions to a certain extent. But we also noted in Chapter 7 the limited extent of these changes. Holton Hall was not rebuilt as a gentry 'seat', and its staff was never headed by a butler. Thomas John may have played the squire to his immediate neighbours, but that did not equate to acceptance into gentry circles. The Dixons' habits had not noticeably altered or their contacts much enlarged in the preceding three decades: no London season or round of shooting parties for them. One historian has claimed that the Dixons were 'a prosperous farming family transformed at mid-century when Thomas John Dixon decided to become a squire'.[5] But I have found no direct evidence for this, and the circumstantial evidence is lacking. Had he made such a decision he would have had to give up farming on a commercial scale; and he would have had also to groom his son for gentrification by giving him a suitable education and finding him a suitable wife. As we know, he did none of these things. He continued the farming business for another decade and a half in the hope that his son would succeed him in it. Richard, it is true, did not receive the hitherto standard grammar school education, but that may have been partly because of his health, and in any case his marriage put an end to any possibility of social advancement

[5] Gerard, *Country House Life,* 14.

through a gentry alliance. Even Thomas John's will was far from a conventional upper-class document, and did not provide for the long-term future of the estate in the way that a strict settlement would have done.

By the mid-1860s it would have been unrealistic to expect the elderly Mr and Mrs Dixon to make any major changes to their routine. Thomas John upheld what he felt to be the dignity of his position, but like his forebears he disliked extravagance and conspicuous consumption. The family tradition of unostentatious living had served it well in the past, and he was unlikely to forsake that tradition now. His hopes for the future of his family, as we have seen, were unfulfilled. Ironically, however, it was following his death, and the extinction of his dynastic hopes, that Holton became a more typical gentry estate. The ladies of Holton may not have been exemplars of cosmopolitan refinement, but they were not engaged in business: they had trustees, solicitors and agents to manage their affairs for them. It was under Ann's will that Holton was settled in a more thorough way, ensuring the succession of its owners into the mid-twentieth century, and under Amelia that the large home farm, the surviving relic of her father's agricultural empire, was finally brought to an end.

* * *

So far this chapter has discussed the position of the Dixons in relation to those above and below them in rural society – in other words, in relation to the rural class system. But in doing so it has been impossible to avoid spatial or geographical references. In the countryside 'knowing one's place' meant knowing how one stood with one's neighbours as well as with one's superiors and inferiors. Class was inextricably connected with community. Thus the squirearchical status of the Dixons was primarily a matter of their relations with those who lived in Holton and on the Holton estate.[6] Their membership of the middling sort depended on their standing in a wider district or neighbourhood. And their acceptance, or lack of acceptance, into the landed gentry was largely a matter of their standing at county level. These three 'social areas' will now be looked at more closely.

The role of the Dixons in Holton itself was affected by the unusual nature of the parochial community, and this in turn was related to the way

---

[6] William Richardson of Limber (d. 1830) was known locally as Squire Richardson although he was a Yarborough tenant and owned no land in the parish (Mary E. Richardson, *The Life of a Great Sportsman (John Maunsell Richardson)* (London, 1919), 4 and *passim*). William's nephew, however, a great horseman, performed the social leap of marrying the widow of the third Earl of Yarborough. His sister Mary's *Life* of him is full of the interesting social nuances arising from this.

in which the family built up its farming and landowning business. In the early nineteenth century it became not only the dominant employer in the parish but also the owner of most of its soil. But this engrossment of power and position came at a cost. First the Bett and Broughton families, once an important part of the social fabric of the parish, disappeared. Then the smallholding tradition, represented by such families as the Hewitts, Nobles and Maddisons, was weakened to the point of extinction; and the close connection between the working families of Holton and their Dixon employers dwindled in importance as Thomas John used more labour from outside the parish. By 1860 he was, to use a military metaphor, the undisputed generalissimo of Holton. But where were the company commanders and the N.C.O.s? The position altered somewhat when Holton was converted into a tenanted estate, and the building of farmhouses at Stope Hill and Ewefield indicated an intention to create a resident tenant-farmer class in the parish. But the new tenants, as we have seen, were not on the whole men of very solid substance, and the onset of agricultural depression meant that they had to concentrate on the survival of their businesses, leaving them little time or money to contribute to the life of the parish. After Thomas John's death the ladies at the Hall exercised their benevolence on the *villagers*, but their personal influence on the parish as a whole was less strong. (The development of the little 'sub-community' around the station has already been noted.[7]) Beyond the parish boundary, in Thornton and Nettleton, the Dixons' influence declined during the nineteenth century. By the 1870s they had become the largest landowners in Nettleton, but the parish retained its 'open' and independent character.

In their heyday the Dixons were at the centre of a network of neighbours, customers and suppliers that extended beyond the three parishes of Holton, Thornton and Nettleton.[8] They had dealings in nearby parishes such as Owersby, Claxby and Normanby, all places, incidentally, with which they had old family connections. But by the late eighteenth century this 'neighbourhood area' was less important to the family socially than the area centred on its market town.[9] Chapter 2 showed how William

---

[7] For country railway stations as social centres, see, for instance, David St John Thomas, *The Country Railway* (Newton Abbot, 1976).

[8] Typical transactions recorded in the Dixon farm accounts were the sale and purchase of small quantities of corn and other produce (which might count as retail rather than wholesale activity in conventional middle-class terms), the loan or hire of men or equipment, the short-term renting of fields and the hiring of animals for breeding purposes.

[9] For the market town area, see also Chapter 1.

senior originated from the Market Rasen district, but how his interests and residence in Holton shifted his allegiance in the direction of Caistor. As a grazier, however, the weekly corn market was of subsidiary importance in his business, and it was not until the last quarter of the eighteenth century that Caistor was firmly established as a commercial and social hub for the Dixon family. Both Thomas and William were by then growing more corn, and were more frequently using the facilities that Caistor, small though it was, could offer as a market town. During the period 1800–30 the Dixons were at their most prominent in Caistor affairs, both fostering and benefiting from the town's access of prosperity. William junior took a lead in local affairs. His three boys, all educated at Caistor Grammar School, became intimately involved in the economy of the town: two of them settled there, while the third, Thomas John, lived only three miles away.

This neat picture must nevertheless be modified in two respects. The affairs of farmers as large as the Dixons could not be confined within the area of a single market town, let alone a town as small as Caistor. They did not send their corn or stock to one market or fair only. For William junior Brigg, with its position on the river Ancholme, provided an alternative market for his corn and the best market for his rabbit skins. And of course the family continued to send its wool to Yorkshire, and occasionally its livestock to London. Secondly, the importance of Caistor in the Dixon family gradually declined after the death of William junior in 1824. The increasing chaos of Marmaduke's affairs was followed by his posthumous bankruptcy. James Green took on the mantle of his father's philanthropic interests, but neither they nor his own corn business flourished. William himself had remained very much the countryman in his attitudes, disliking the gaieties of Caistor's social life; and Thomas John, although maintaining his links with the town, played only a limited role in its public affairs. He spread his commercial contacts into Yorkshire, Nottinghamshire and Lancashire, and locally his Searby interests tended to increase his links with Brigg rather than with Caistor. The latter's economy, moreover, declined during Thomas John's time, hastened by the arrival of the railways, or rather their failure to come within more than a few miles of the town. Only under Mrs Jameson Dixon, when the Dixons' world had closed in on itself, did Caistor's links with Holton revive. She married a Caistor doctor; she lived in Caistor itself for some years; and when she took up residence as lady of the manor of Holton she brought her Caistor servants with her.[10]

---

[10] For Caistor's nineteenth-century doctors, including Dr Jameson, see David Saunders, *More Portraits of Caistor, Lincolnshire* (Heighington, 2007), 5–43.

If the Caistor market area was too small to contain the Dixons, was there a larger district of north Lincolnshire into which they fitted more comfortably? One candidate already mentioned in these pages was the Lincolnshire Wolds. William senior had grazed them at Normanby and Claxby; Thomas had settled on their eastern edge; Thomas John farmed wold land at Nettleton and Searby, as did James Green at Rothwell. Both Thomas John and James Green had dealings with many of the wold farming families, and Thomas John named representatives of two of them as trustees in his will. But these links should not be exaggerated. William senior ceased to farm on the Wolds in later life. Thomas's connections at Riby were more with lowland parishes such as Laceby and Healing than with the uplands (with the exception of the family connection at Limber). William junior had little in common with the hard drinking and hard riding side of farming culture on the Wolds, and his writings betray no consciousness of the Wolds themselves as a distinct country or *pays*. Thomas John rode and even hunted as a young man, but he joined the Rasen rather than the Yarborough yeomanry troop,[11] and in later years his personal dealings with the wold farmers decreased.

It was a similar story with the Dixons' relations with the Anderson-Pelhams, Lords Yarborough, whose influence stretched across much of the northern Wolds. Thomas Dixon was a recipient of Brocklesby patronage: Charles Anderson-Pelham helped to make him a J.P. and gave his son Thomas a living. But these links were not maintained in the next generation. William junior's marriage connected him only remotely with Brocklesby: the Ravendale Parkinsons were close to Brocklesby, but apparently not close to their cousins at Holton. William himself, as we have seen, had a somewhat ambivalent attitude to Lord Yarborough, and regretted his lack of whole-hearted support for the Caistor Society of Industry. Thomas John co-operated with Lord Worsley, later the second Earl, in railway matters, gave one of his votes to the Yarborough candidate at elections and received the occasional invitation to Brocklesby, but was at best a semi-detached supporter.

For William junior it can be argued instead that the district that meant most to him in terms of local community was indeed based on Caistor, but was wider than the Caistor market area. In fact it had a radius more like ten miles than the four or so of Caistor's commercial hinterland. This wider district took in part of the Wolds but also adjacent lowland areas

---

[11] The captain of the Market Rasen troop, Ayscoughe Boucherett, did however have a Caistor connection as hereditary patron of the grammar school.

in the Ancholme valley. It embraced both the Holton and the Riby neighbourhoods, included Brigg and Market Rasen as well as Caistor, and, most significantly, corresponded closely with the area covered by the united parishes of the Society of Industry.[12]

This district, however, lost some of its coherence in Thomas John's time. The area covered by the Caistor union was enlarged under the New Poor Law. The family connections with the Riby, Healing and Barnoldby area faded away. The family farms became concentrated in an area more closely corresponding to the Brigg and Caistor market districts. This concentration within a relatively small area seems to have been untypical of the most prominent north Lincolnshire farming families at that period. The Richardsons of Limber, for instance, were established in the Kirton-in-Lindsey as well as the Grimsby neighbourhoods. The Brooks family had land or farms in the neighbourhoods of Barton and Grimsby as well as Caistor. On a larger scale the Skipworths certainly concentrated themselves on South Kelsey, but they had outliers at Aylesby, near Grimsby, Legbourne, near Louth, and Nettleham, near Lincoln.[13]

As noted in Chapter 1, Lincoln was some distance from Holton, and not linked easily to it by decent roads. It was of course the shire town, and members of the family journeyed to it from time to time to attend county meetings or to sit on the Grand Jury at the Assizes. They might, too, be drawn occasionally to the city on diocesan business. But, at least before 1850, Lincoln exerted no strong commercial pull on the environs of Holton, and it was not (before 1889) an administrative centre for the county division of Lindsey. It is not therefore very surprising that Thomas and William junior confined their public service for the most part to their own locality. Thomas John, however, was a little different. Though no committee man or public speaker, he was seen more frequently in Lincoln than previous members of the family, his journeys facilitated by the railway from 1848. Like his grandfather he sat on the Grand Jury. His name began to appear more frequently on county-wide subscription lists, and in

---

[12] See also William's own definition of a community as containing more than one market town, in Chapter 5 above.

[13] In a study of south-east Surrey Evelyn Lord emphasises the importance of family 'dynasties' within the farming class, and describes these as often extending over community boundaries. One family's 'social area' might cover two to four local communities, and the family might also have links with larger but more distant population centres (in this case London and Croydon). ('Communities of common interest: the social landscape of south-east Surrey 1750–1850', in Charles Phythian-Adams, ed., *Societies, Cultures and Kinship 1580–1850: Cultural Provinces and English Local History* (Leicester, 1993), 137–99.)

1862 he achieved undeniable county status as High Sheriff, a few years after his brother-in-law George Skipworth. But the appointment was for one year only, and came too late to widen his social connections. More significant over the years were his links with Yorkshire, strengthened through his marriage and, again, facilitated by the ease of railway travel. These shifts continued after his death. Towards the end of the century the Holton trusteeship business was moved from Caistor to Lincoln; but the estate business was effectively transferred to Yorkshire, where it was run by a solicitor in Beverley and an agent based in Hull.

* * *

Chapter 9 established that very few north Lincolnshire farming families acquired large landed estates in that division of the county in the century or so before 1870. It was rare to purchase over 1,000 acres of land, and even rarer to accumulate an estate of over 2,000. Various economic reasons were adduced for this phenomenon, relating to factors of both supply and demand. What has been said in this chapter, however, suggests that there could also be social reasons. The large-scale ownership of the soil was intimately connected with the structure of rural society, and the purchase of an estate, whether by a local farmer or a businessman from outside the county, could be interpreted as a challenge to the social order.

It was a challenge that might not worry the magnate on a county-wide scale, such as Lord Willoughby de Eresby or Lord Yarborough. But lesser gentry, conscious perhaps that their ancient lineage was not matched by the breadth of their acres, might take it more seriously, and it was often they who had the power to influence opinion at a more local level.[14] And it was not just the gentry who adhered to hierarchical notions of society. William Dixon junior was himself a purchaser of land, but his acquisitions were incremental, and on a modest scale – a different matter, in his view, from the very conspicuous purchase of the South Kelsey estate by the Skipworth family, which he saw as subversive of the social order. Middle-class people, even if they had ceased to engage in the businesses through which they had made their money, found it very difficult to gain acceptance into the ranks of the landed gentry during the nineteenth century, and indeed for some decades after that.[15] One could find oneself in a kind of social

---

[14] For one such squire, Sir Charles Anderson of Lea, see Olney, *Rural Society*, 40–1.

[15] For the twentieth century, see Howard Newby, Colin Bell, David Rose and Peter Saunders, *Property, Paternalism and Power: Class and Conflict in Rural England* (London, 1978), 299–306.

limbo, stranded between the middle and the upper class, and this is arguably what happened to Thomas John Dixon in the latter part of his life.

There was another aspect to the social implications of landownership, and that was the relationship between social and geographical mobility.[16] If a middle-class family wished to elevate itself into the landed gentry, it made sense to make the attempt at some distance from the district in which its fortune had been made. Perhaps such a thought had been in the mind of Thomas John's London lawyer when he suggested purchasing an estate in Herefordshire. It was certainly more difficult to change one's class while remaining in one's native district. When Charles Tennyson d'Eyncourt built his Gothic castle at Tealby local people could still remember his father as a junior in the attorney's office in Market Rasen, only three miles away.[17]

Farmers who wanted to buy land in any quantity had particular problems in choosing *where* to buy it. As discussed in Chapter 9, it was often difficult enough to acquire suitable nearby land to add to one's holding, without the added problems of the effects on local society. The most attractive solution was to buy one's own farm from one's landlord. If the estate were an outlying one from the landlord's point of view, and this was often the case, the social disruption would be minimal. The farmer's relations with his workpeople would not alter, since he may already have been acting as a stand-in squire. But such opportunities were rare. If, on the other hand, he bought land at some distance from his place of residence there would be obvious problems of supervision or management, even if the estate were bought for investment rather than for farming purposes.

Chapter 9 does not discuss those north Lincolnshire farmers who bought land outside the division, but in fact very few seem to have done so.[18] The inconvenience involved in purely business terms may have been one factor, but another may have been the preference of farmers for remaining within the local communities to which they belonged, and with whose workings they were familiar. To intrude into another neighbourhood, moreover, might excite the resentment of older-established local farming families. We may recall that when Philip Skipworth became a major landowner the challenge to the social order was not his only fault

[16] Colin R. Bell, *Middle Class Families* (London, 1968), 95ff.
[17] Olney, *Rural Society*, 40–2.
[18] A rare example was J.W. Dudding of Saxby, near Lincoln, the son of the noted sheep-breeder (and himself sometime trustee for the Searby estate), who purchased a farm at Howell in Kesteven in 1863. I know of no north Lincolnshire farmer who purchased a significant amount of land outside Lincolnshire altogether.

in William Dixon's eyes. He also offended against the conventions of the local community, despite the fact that he was not exactly a stranger to it. Elsewhere, however, in more sparsely populated parts of the county where there was less sense of community, the purchase of a parish of 1,000 or 1,500 acres may have caused less unfavourable comment; and in any case families such as the Sowerbys and Coateses, with comparatively shallow roots in north Lincolnshire, would no doubt have taken a few grumbles in their stride. Such farmers can be seen as the forerunners of a more numerous race of farmer-landowners who were to become prominent in north Lincolnshire in the twentieth century, when many of the aristocratic estates were reduced or broken up.

* * *

In short, the average north Lincolnshire farmer of the eighteenth and nineteenth centuries, however successful in his business, and however devoted to increasing his 'proputty', was essentially conservative in his social attitudes. He accepted his position in the rural hierarchy, and within his local community or communities; and the thought of moving into uncharted social, or indeed geographical, territory, was one to give him pause. All this makes the Dixons' forays into the purchase of land – with some reluctance in the case of William junior, with more energy in the case of William senior and Thomas John – the more remarkable. Even within a tiny cohort of similar families they were untypical, in rising to prominence both before and after rather than during the Napoleonic period, and by remaining actively involved in farming long after most of their landed neighbours.

It could be argued that this untypicality reduces their interest for the historian. That might perhaps be true if they could not be studied in their local context, but in fact it has been possible to show them as thoroughly embedded in the network of kinship, commercial and social connection that defined their neighbourhood. Successive heads of the Dixon family differed in their personalities, their achievements and the part that they played in local affairs. But as a *family* their history sheds light on the *mores* of the rural middling sort – not just on their attitudes to landownership but on their family structures and alliances; their attitudes to education and the transmission of wealth from one generation to another; the important part played by their womenfolk; and the ways in which they defined their role in the parish, the market area and the county at large.

Although the Dixons did not add to their landed portfolio during the years 1783 to 1815, that period can nevertheless be seen as a kind of high-water mark for the middling sort of north Lincolnshire. They were certainly years when the Dixons themselves were at the height of their local prominence and influence. Thereafter it is possible to detect a loosening of local ties and a weakening of local networks, as new means of transport opened up the countryside, and as national ideas of class gained over local ideas of community. It is also possible to see a reflection of these changes in the declining commercial and social importance of the little market town of Caistor, with which the fortunes of the Dixon family were so closely involved.

* * *

The story of Holton-le-Moor and its inhabitants does not of course end in 1906. In some ways the twentieth-century history of the village and parish is just as remarkable as that of the eighteenth and nineteenth. There is, for instance, the work of T.G. Dixon in reviving his adopted local community in the early decades of the century,[19] and, towards the end of the century, the reconversion of the Holton estate to a farming enterprise in some ways reminiscent of the early Victorian period. But these are topics for the historian's future attention.

---

[19] For which, however, see Olney, 'Squire and community: T.G. Dixon at Holton-le-Moor 1906–1937', *Rural History* 18 (2007), 201–16; Richard Olney and Barry Graham-Rack, *The Moot Hall, Holton-le-Moor: 100 Years, 1910–2010* (printed, 2010).

APPENDIX 1
THE DIXON ARCHIVE

This book is based principally on a rich collection of family papers, accumulated by the Dixons over several generations. Indeed, had it not been for the existence of this collection it would have been impossible to write the book in this form. Such collections of estate and family papers are an important feature of English archival history, and their survival has been due above all to two related factors – the establishment of family traditions whereby records have been created, maintained and preserved over long periods, and the continuous existence of a safe and stable environment in which this process can take place. The Dixon archive meets both these criteria. The family has been good at creating and keeping records since the time of Thomas Dixon (1729–1798). And since 1785 the family has owned and lived in one house, the Hall at Holton-le-Moor, where a good proportion of the records have been concentrated, although since the 1970s the older part of the archive has been on deposit at Lincolnshire Archives. (See below, Principal Sources.) The collection has never been the victim of a failure of heirs or a dispersal by sale, nor has it suffered from the physical effects of fire, flood, neglect or removal from place to place, all hazards that have caused loss or damage to many other collections of family papers.

In many respects the Dixon papers fall into a number of standard archival categories – deeds and legal papers, including papers relating to wills and family settlements; estate records, including maps, surveys, rentals and accounts; papers relating to office-holding and local affairs; and personal papers, including letters and diaries. But in three respects the collection is unusual. Unlike many estate collections, it includes detailed series of farming accounts and papers. For one member of the family in particular, William Dixon (1756–1824), there survive some remarkable personal papers, a rarity for working farmers of that period. And in the twentieth century T.G. Dixon and his son G.S. Dixon enriched the collection with a number of items and groups of papers from other sources.

The deeds and legal papers provide a clear illustration of the importance of Holton Hall as a repository for the family archive. There is in fact a shortage of deeds for the period before the building and occupation of the Hall: the earlier title, for instance, to the manorial estate of Holton itself consists mainly of copies. Before the early nineteenth century the family appears to have used a firm of solicitors at Barton-on-Humber, of whose holdings for that period only a few fragments appear to have survived. From 1810, however, Marmaduke Dixon had his own practice at Caistor, and until his death in 1830 he handled much of the family's legal business, although his brother Thomas John was also keeping his own careful records at Holton. Following Marmaduke's death his practice was taken over by Marris and Smith, but Thomas John increasingly preferred to keep his deeds and related papers directly under his eye at Holton. The alterations to the Hall in 1839–40 gave him an opportunity to set up a new business room at the back of the house, and his mother-in-law Mrs Roadley met the cost of a Chubb safe for it.[1] The tradition of keeping the main series of deed bundles at Holton continued into the twentieth century. T.G. Dixon moved the records (and the safe) into a newly built study behind the drawing room shortly before the First World War.

Despite this concentration at Holton, however, various caches of legal papers remained with the firms of solicitors that acted for the family. Thomas John continued to use the local Caistor firm of Marris and Smith, but from time to time he also consulted the London firm of Eyre and Coverdale, who acted as the London agents for Marris and Smith (and before them for Marmaduke Dixon). The Roadleys relied on the Beverley firm of Shepherd and Myers, later Crust, Todd and Mills, although once Thomas John became involved in the Roadleys' affairs through his wife he began to accumulate related papers. In the late nineteenth century the Dixon trust business was handled by the Lincoln firm of Hebb and Sills, but Mrs Dixon and her daughters continued to use Crust, Todd and Mills for their personal business. Eyre and Coverdale also acted for the Skipworth family.[2]

Central to the Dixons' attitude to record-keeping was their fondness for – almost their obsession with – accounts. Regular and careful accounting was deemed essential to the efficient management of their business, and to their reputation for fair dealing with their neighbours, but it was also, at least with William Dixon (1756–1824), a matter of accounting to

---

[1] DIXON 9/1/38; 8/1, pp. 30ff.

[2] Dixon trust papers seen at the Lincoln office of Sills and Betteridge (successors to Hebb and Sills) in 1970 were later transferred to Messrs Bates and Mountain of Caistor. A few Dixon and Skipworth items were seen at Collyer Bristow (successors to Eyre and Coverdale) in 1990. (For a transfer of Skipworth trust records from Collyer Bristow to Holton in 1912, see below.) An attempt to trace clients' papers for the Roadley family at Beverley was unsuccessful.

God for the proper use of one's talents. William Dixon senior (1697–1781) must have been literate and numerate, but none of his accounts or personal papers survives. His son Thomas, however, preserved his papers: in particular his general ledger, covering the period 1755–98, set a precedent that was followed by successive members of the family. He died, however, not at Holton but at his rented farm-house at Riby. We do not know what was destroyed there, but the papers that do survive were presumably taken to Holton by his son William.[3]

It was the nineteenth century that saw the hey-day of financial record-keeping at Holton. The young Thomas John Dixon kept detailed accounts from the earliest years of the century, encouraged no doubt first by his father and then by his uncle and partner Richard Roadley; but the surviving records are most impressive from 1823–4, when he completed the takeover of the Holton business from his father. The general ledgers remained the key to the accounting system, but they were supported most notably by a fine series of day books, a mixture of labour and cash accounts that run continuously from 1824 to the final demise of the home farm in 1897. Long before 1897, however, the Holton enterprise had been evolving from a group of farms into an agricultural estate. As a consequence the day books document the increasing involvement of bailiffs and agents, although during his lifetime the guiding hand remained that of Thomas John himself. A number of subsidiary accounts have also survived, dealing with corn, seeds, stock, wool and labour. Of these it is the labour accounts that are of most interest from the archival point of view, since they incorporate records created by some of the labourers themselves, workers who had been taught the rudiments of letters and figures in the Holton Sunday school. The value placed on all these records by Thomas John himself is shown by the provision in his will that the account books should pass as heirlooms with the estate.

As for papers relating to local affairs, the justice books of Thomas Dixon are a rare survival for the late eighteenth century.[4] As far as personal papers are concerned, however, the outstanding contributor to the Holton archive was his son William. Like other members of the family he wrote letters and carried about with him a small pocket book that functioned as a

---

[3] We know that certain items from Riby (DIXON 8/1, 22/8/1) were at Holton in the mid-nineteenth century, since they were reused at that period by Thomas John Dixon. (Reusing volumes and blank sheets for other purposes was a characteristic of Holton record-keeping.) But Thomas's big ledger is believed to have passed to J.G. Dixon and to have reached Holton only in 1918. (For J.G. Dixon's papers, see also below.)

[4] See Lincoln Record Society, 102 (2012).

diary and memorandum book. But he also compiled the series of revealing jottings and reflections, most fully preserved for the period 1805–24, that are discussed in Chapter 5.

Following the death of Thomas John Dixon in 1871 the legal and farm-cum-estate records that had accumulated in the house down to that point became to a certain extent a closed archive. Legal papers were now with the solicitors, and accounts and papers relating to the Holton farms and estate were kept mainly by the accountant and sub-agent William Lord, who had an office behind the house that he occupied until his death in 1908. Mrs Dixon and her two daughters did keep their own papers relating to the running of the household and to their private affairs, but there is a dearth of surviving papers for Ann Dixon, the result, it is believed, of destruction after her death by her sister and successor Amelia Margaretta.

Nevertheless when the Gibbons family arrived at Holton in 1907 they found a veritable Aladdin's cave of records. T.G. (Tom) Dixon set about sifting through all the material that he could find in the house that related to the place and the family and, from 1913–14, adding it to the collection of deeds and other papers in the study. He also obtained groups of material from elsewhere, most significantly Skipworth papers from the office of Collyer Bristow, the successors to Eyre and Coverdale, in 1912, and papers of James Green Dixon, Thomas John's younger farming brother, and his family, retrieved from a house in Caistor after the death of J.G. Dixon junior in 1918. Tom Dixon's son George, a keen antiquary and genealogist, acquired further groups of papers by gift, purchase or loan, among them parts of the Parkinson and Tennyson d'Eyncourt family archives, and the pedigree books of John Kirkby of Caistor (d. 1923), a painstaking genealogist whose special interest in north Lincolnshire farming families have been a valuable source for this book.

George Dixon was a friend of the Lincolnshire Archives Office, and, particularly from the late 1950s, made various deposits of deeds and papers in it.[5] But they were not substantial, and the full extent of the manuscript riches at Holton was revealed only near the end of his life, when he decided to deposit the archive almost in its entirety.[6] The work of sorting the papers and transferring them to the Archives Office began in 1969,

---

[5] Lincolnshire Archives Committee, *Archivists' Reports* 11 (1959–60), 49–50 (MISC DON 140); 13 (1961–2), 49; 14 (1962–3), 49; 15 (1963–4), 23; 16 (1964–5), 24; 18 (1966–7), 59–60. For an earlier deposit of Tennyson d'Eyncourt MSS, see also LIND DEP 106. For a deposit through the Lincolnshire Architectural and Archaeological Society, see AS 4B.

[6] *Archivists' Report* 22 (1970–1), 17; personal knowledge.

and continued after his death in 1970, since his sisters needed to clear the house before handing it over to the heir. My colleague Michael Lloyd and I made a number of visits to Holton Hall, where our work was concentrated in the study and the loft above it. One chair in the study held layers of newspaper cuttings, which the maids had always been told not to disturb: dealing with it was as much an archaeological as an archival exercise. The loft, into which George had been unable to climb in his latter years, was another problem: we had to lower manuscripts from it in baskets to the study floor below. The discoveries continued. In 1972 a locked safe in the old estate office yielded more nineteenth-century ledgers, and a whole separate family and estate archive, that of George's mother's family of Sperling, was despatched from the long granary to Essex Record Office.[7]

Back at the Archives Office it was decided to begin listing the collection in December 1970, with subsequent deposits being added to the catalogue as they arrived. The list also incorporated the small deposits made prior to 1969. The main collection, DIXON, was treated as a deposit from the owner of the estate for the time being.[8] George had, however, bequeathed his personal correspondence and papers to the Archives Office, and these were listed as 2 DIXON. He had also generously left the Office such books from his library as it wished to take: these included works of reference that filled gaps in its already excellent library, and a number of volumes acquired by, and sometimes annotated by, members of the family in the eighteenth and nineteenth centuries.[9] The close relations between Holton and the Archives Office were maintained by the new owner of Holton, Mr Philip Gibbons, who continued to send papers to Lincoln to join the main deposit.[10]

One or two items disappeared from the Holton archive before the 1970s, despite the vigilant care of its custodians. They may have included a volume relating to the Caistor Canal and the first log book of Holton

---

[7] Essex Record Office, Sperling Papers (D/Dgd). This collection includes papers of the Revd T.G. Dixon, supplementary to those deposited at Lincoln, and even a few strays from the Tennyson d'Eyncourt papers. The study and the long granary at Holton were both later demolished.
[8] *Archivists' Reports* 22 (1970–1), 17–24; 23 (1971–2), 9–10; 24 (1972–3), 20–2. The later additions to DIXON classes 1–12 are catalogued as DIXON 22. For the sections of the archive most frequently used in this volume, see below, Principal Sources.
[9] For 2 DIXON and the Dixon Library, see *Archivists' Report* 23 (1971–2), 10–11. Additional papers incorporated in 2 DIXON were deposited by the Misses Gibbons in 1979–80 (*Archivists' Report* 27 (1977–82), 34).
[10] 3–5 DIXON, deposited 1975–2002. A deposit by Miss Dora Gibbons from Haycorns, Holton-le-Moor, in 1986 was catalogued as MISC DEP 487.

School, perhaps loans that were not returned. But, needless to say, what survives is of incomparably greater value than what has been lost. At one time related collections of family papers would have been kept by the Roadleys at Searby Manor and by the Skipworths at Moortown House, but Tom Dixon arrived too late on the north Lincolnshire scene – a few years too late in the case of Moortown House, several decades in the case of Searby Manor – to salvage anything for the collection at Holton. Again, however, these losses serve as a reminder of the remarkable care and good fortune that has preserved the Dixon archive itself.

APPENDIX 2
GENEALOGICAL TABLES

TABLE I: DIXON

Robert of Owersby (d.1593)
|
Robert of Usselby (1570-1627)
|
William of Middle Rasen (1603-1662)
|
Thomas of Kirkby-cum-Osgodby, later of Market Rasen (d.1704),
m. (3rdly) Ann Waltham of Wragby (she m. 2ndly Sailbanks Broughton
of Market Rasen, later of Holton-le-Moor)
|
William of Holton-le-Moor (1697-1781)
m. *c.*1726 Rachel Drewry of Adlingfleet (Yorks.)
|
Thomas of Riby (1729-1798)
m. 1755 Martha Walkden of Great Limber
|
———
| | | | | | |
Rachel (b.1757) Revd Thomas   Revd Richard   Martha   Mary   Jane   Marmaduke   Ann (1772-1858)
m. William Etherington of Laceby   of Claxby-by-Normanby (1762-1783) (b.1764)   (b.1766)   (b.1768)   m.1798 Richard **Roadley**
of Gainsborough   (1759-1832)   (1760-1819)   m. R. Smith of   m. J. Kelk
                                                             Skipton (Yorks.) of Brigg

William Dixon of Holton-le-Moor (1756-1824)
m. 1782 Amelia Margaretta **Parkinson**
|

Family Tree

Generation 1 (siblings):
- *Thomas John* (1785-1871) m.1827 Mary Ann **Roadley**
- Frances Martha (1786-1810)
- Amelia Margaretta (1787-1793)
- Marmaduke (1788-1830) m.1814 Susanna Atkinson
- Rachel Harriott (d.1791)
- William (d.1793)
- Amelia Margaretta (1794-1879) m.1815 George **Skipworth**
- James Green (1789-1879) m.1825 Elizabeth Dauber

Children of Thomas John & Mary Ann Roadley:
- *Ann* (1828-1893)
- William (1829-1830)
- *Richard Roadley* (1830-1871) m.1860 Lucy Collinson

Children of Marmaduke & Susanna Atkinson:
- *Amelia Margaretta* (1833-1906) m.1876 Dr George Jameson
- Thomas John (1835-1855)
- Charlotte Roadley (1836-1854)

Children of Amelia Margaretta & George Skipworth:
- Mary Ann (1839-1856)

Children of James Green & Elizabeth Dauber:
- Mary Elizabeth (1836-1875)

Children of Richard Roadley & Lucy Collinson:
- John William (1826-1898) m. Ellen Josephine Peet and had issue
- Thomas Parkinson
- James Green (1833-1918)
- Marmaduke (1828-1895) m.1859 Eliza Agnes Wood and had issue

193

TABLE II: PARKINSON

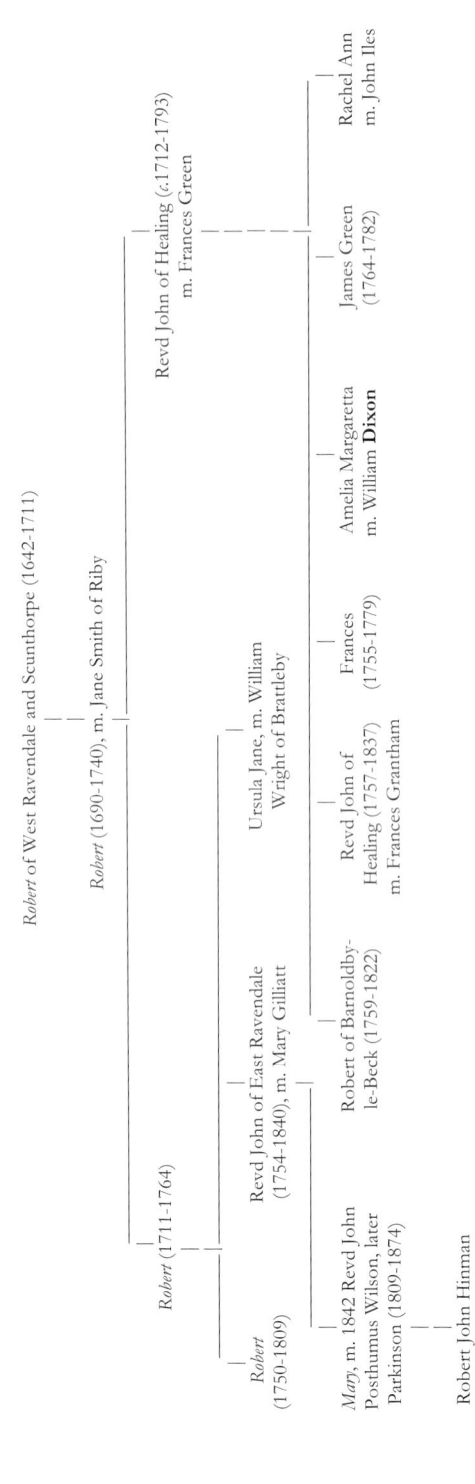

Robert of West Ravendale and Scunthorpe (1642-1711)

Robert (1690-1740), m. Jane Smith of Riby

- *Robert* (1711-1764)
  - Robert (1750-1809)
  - Revd John of East Ravendale (1754-1840), m. Mary Gilliatt
    - Mary, m. 1842 Revd John Posthumus Wilson, later Parkinson (1809-1874)
      - Robert John Hinman Parkinson (b.1844)
  - Robert of Barnoldby-le-Beck (1759-1822)
  - Ursula Jane, m. William Wright of Bratleby
  - Revd John of Healing (1757-1837), m. Frances Grantham
  - Frances (1755-1779)
- Revd John of Healing (c.1712-1793), m. Frances Green
  - Amelia Margaretta m. William **Dixon**
  - James Green (1764-1782)
  - Rachel Ann m. John Iles

TABLE III: ROADLEY

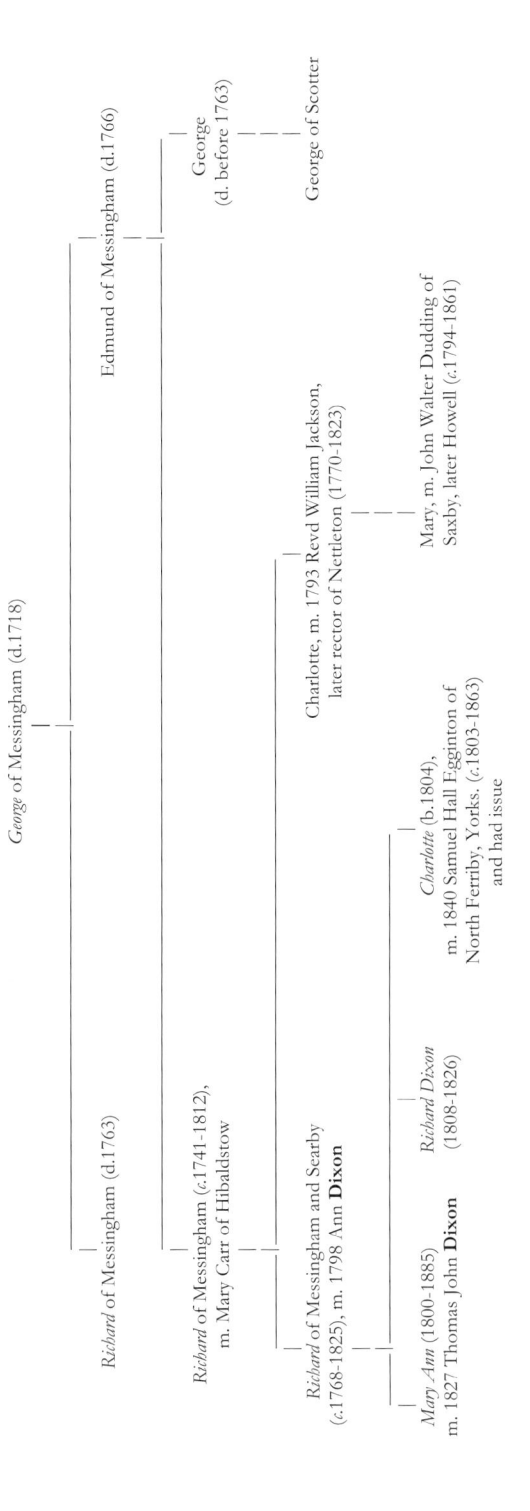

```
                              George of Messingham (d.1718)
                                         |
              ┌──────────────────────────┴──────────────────────┐
    Richard of Messingham (d.1763)                    Edmund of Messingham (d.1766)
              |                                                 |
                                                    ┌───────────┴───────────┐
    Richard of Messingham (c.1741-1812),         George                 George of Scotter
    m. Mary Carr of Hibaldstow                (d. before 1763)
              |
    ┌─────────┴──────────────────────────┬─────────────────────────────┐
  Richard of Messingham and Searby    Charlotte, m. 1793 Revd William Jackson,
  (c.1768-1825), m. 1798 Ann Dixon      later rector of Nettleton (1770-1823)
              |                                     |
    ┌─────────┼──────────────┐         Mary, m. John Walter Dudding of
  Mary Ann (1800-1885)  Richard Dixon    Saxby, later Howell (c.1794-1861)
  m. 1827 Thomas John Dixon  (1808-1826)
                         Charlotte (b.1804),
                         m. 1840 Samuel Hall Egginton of
                         North Ferriby, Yorks. (c.1803-1863)
                         and had issue
```

195

TABLE IV: SKIPWORTH

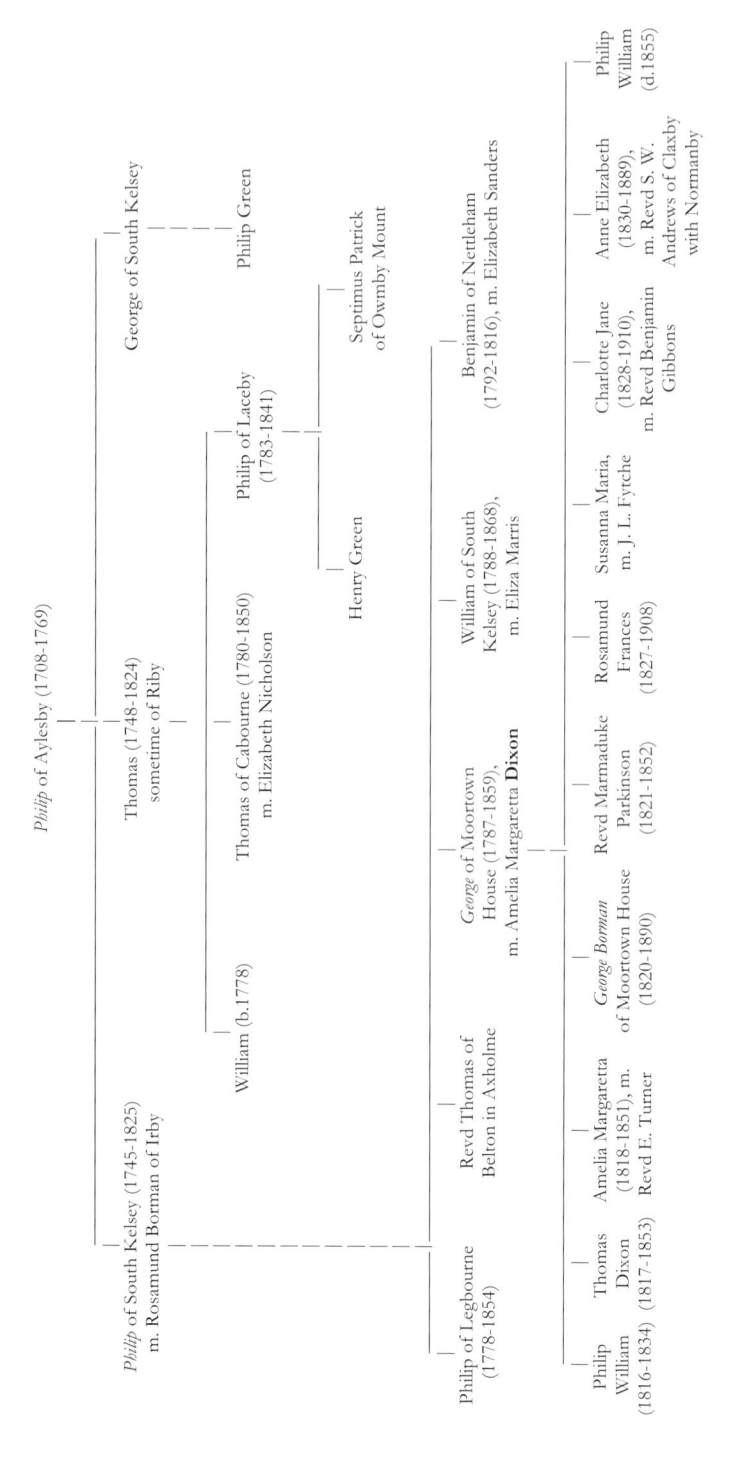

Philip of Aylesby (1708-1769)
│
├── *Philip* of South Kelsey (1745-1825), m. Rosamund Borman of Irby
│   │
│   ├── Philip of Legbourne (1778-1854)
│   │
│   ├── Revd Thomas of Belton in Axholme
│   │
│   └── *George* of Moortown House (1787-1859), m. Amelia Margaretta **Dixon**
│   │
│   ├── Philip William (1816-1834)
│   ├── Thomas Dixon (1817-1853)
│   ├── Amelia Margaretta (1818-1851), m. Revd E. Turner
│   ├── *George Borman* of Moortown House (1820-1890)
│   └── Revd Marmaduke Parkinson (1821-1852)
│
├── Thomas (1748-1824) sometime of Riby
│   │
│   ├── William (b.1778)
│   │
│   ├── Thomas of Cabourne (1780-1850), m. Elizabeth Nicholson
│   │
│   ├── Philip of Laceby (1783-1841)
│   │   │
│   │   └── Henry Green
│   │
│   ├── Septimus Patrick of Owmby Mount
│   │
│   └── William of South Kelsey (1788-1868), m. Eliza Marris
│   │
│   ├── Rosamund Frances (1827-1908)
│   ├── Susanna Maria, m. J. L. Fytche
│   ├── Charlotte Jane (1828-1910), m. Revd Benjamin Gibbons
│   ├── Anne Elizabeth (1830-1889), m. Revd S. W. Andrews of Claxby with Normanby
│   └── Philip William (d.1855)
│
├── George of South Kelsey
│
├── Philip Green
│
└── Benjamin of Nettleham (1792-1816), m. Elizabeth Sanders

PRINCIPAL SOURCES

Manuscript sources

Dixon Papers deposited in Lincolnshire Archives from Holton-le-Moor, including

Deeds for Holton, etc (DIXON 1/A–E, 22/1; 3 DIXON 1; 4 DIXON)
Wills, settlements, legal papers and executorship papers, Dixon family (DIXON 2–3, 22/2–3; 3 DIXON 5/11–12; 5 DIXON 3)
Farming and estate records, Holton and Searby (DIXON 5–6, 22/6–7; 3 DIXON 2, 4; 5 DIXON 1–2, 8)
Household records (DIXON 4/18, 22/5)
Personal and financial papers, Thomas Dixon (DIXON 4/1–2, 8, 22/8/1), William Dixon junior (DIXON 4/3–5, 7, 22/8/2–5; 3 DIXON 5/1–2, 5/11/7), Thomas John Dixon and family (DIXON 4/6–19, 22/4; DIXON 9–10, 22/9; 3 DIXON 5/3–11), the Revd T.G. Dixon (DIXON 11, 22/10; 3 DIXON 6), and G.S. Dixon (2 DIXON, 3 DIXON 7)
Papers of James Green Dixon and family (DIXON 12, 22/11)
Papers of the Skipworth family (DIXON 1/F–J, 2/4, 3/8, 13; 5 DIXON 4–7)
Wills and legal papers relating to the Roadley family (DIXON 2/3, 3/7)
Papers of the Parkinson family (DIXON 2/5, 3/9, 16)
Kirkby pedigrees (DIXON 15)
Official records (DIXON 17)
Maps, plans, drawings and photographs (DIXON 18, 21; 3 DIXON 8)

Lincolnshire Archives, other collections

Barton Parish Deeds (BPD): Dixon deeds 18th cent. (apparently strays from the offices of a Barton solicitor, see DIXON 15/1, vol. 38, p. 38)
Messrs Bates and Mountain (of Grimsby, Caistor and Market Rasen, solicitors) (BM): Young of Kingerby deeds (4 BM 4/1–2)
Brownlow papers (BNL): papers of the 1st Earl Brownlow as Lord Lieutenant of Lincolnshire (4 BNL)
Collyer Bristow (of London, solicitors) (BRA 1489): Skipworth trust papers, deposited through the British Records Association
Farrers (of London, solicitors) (BRA 1216): Goulton papers
Foster Library (FL): Garthorpe and district deeds, including Drury (Adlingfleet) deeds (FL Garthorpe and district deeds 1)
Mrs Lamb (LAMB): papers relating to Caistor and district (deriving from the solicitors' practice of Messrs Marris and Smith and their successors), deposited through the good offices of Mr C.J. Ollard

Lincoln Dean and Chapter Records, including leases and probate records
Lincoln Diocesan Records, including bishops' registers 38 and 39, LCC wills and inventories and Holton-le-Moor tithe award (F101)
Lindsey Muniment Room deposits (LIND DEP), including Tomline (Riby) estate papers (LIND DEP 29/5)
Lindsey Quarter Sessions Records (LQS), including Quarter Sessions minutes, Commission of the Peace correspondence, land tax records and enclosure awards
Monson papers (MON), including papers relating to Owersby and North Carlton (MON 9/2B, 25/13/4)
Pretyman-Tomline papers (PT, 2PT), including Riby estate papers (2 PT 2/18, 3/8–13), deposited by Farrers
Redbourne Hall papers (RED): papers relating to the Prebend of Caistor (RED 2/4)
Riby parish records, including registers and churchwardens' accounts
Sills and Betteridge (of Lincoln, solicitors) (SB): deeds relating to the Foster (Barnetby-le-Wold, etc) estate (SB/Foster)
Taylor, Glover and Hill (of Epworth, solicitors) (TGH): papers relating to Claxby-by-Normanby Rectory (2TGH 3/A/4/2), the Collinson family of the Isle of Axholme (2TGH 1/16/2) and the Farrow family of Swallow (2TGH 1/25/5)
Tennyson d'Eyncourt papers (TDE), including papers deposited via Holton (4TDE)
Toynbee, Larken and Evans (of Lincoln, solicitors) (TLE): papers relating to the Slater family of North Carlton (TLE 40)
Whichcote papers (ASW): papers relating to the Meres family and land in Holton-le-Moor (ASW 1B/77) (and see also BRA 1597)
Wright of Brattleby papers (MISC DEP 118), deposited by Mr C.J. Ollard
Yarborough papers (YARB), including Brocklesby estate rentals and accounts (YARB 5)

Collections in other repositories

The National Archives (TNA): census enumerators' returns; Chancery Masters' Exhibits, including papers relating to Marmaduke Dixon (C33/812, etc, C123/I/4–16)
Centre for Buckinghamshire Studies: Carrington papers, including papers relating to the Humberston family's estates in north Lincolnshire (Box 33)
Cambridge University Library: Ely Diocesan Records, including leases for Thornton-le-Moor (CC Bp 94511)
East Riding of Yorkshire Archives and Local Studies Service: Chichester-Constable papers (DDCC), including papers relating to Claxby (DDCC 151)
Herefordshire Record Office: Lee-Warner (of Tiberton) papers, including papers relating to Parkinson (later Dixon) property (K8/80)
North East Lincolnshire Archives: Parkinson (of Ravendale) papers (361, 542/2, 5); Maw (of Cleatham) papers (379/A28, 34)
Nottingham University Library: Middleton papers (Mi), including Rothwell (of Thorganby) papers (Mi Da 67, 73, E 17)
Surrey History Centre: Frederick papers (183), including deeds for Holton

Printed Sources

Specific to Lincolnshire

Beastall, T.W., *The Agricultural Revolution in Lincolnshire* (Lincoln, 1978)

Collins, Jean, *South Kelsey: The History of a North Lincolnshire Village* (Peterborough, 2009)

Davey, B.D. (ed.), *The Justice Books of Thomas Dixon of Riby 1787–1798*, in *The Country Justice and the Blackamoor's Head: The Practice of the Law in Lincolnshire 1787–1838* (Lincoln Record Society 102, 2012)

Dixon, William, *Reports of the Several Institutions of the Society of Industry, established at Caistor A.D. 1800, for the better relief and employment of the poor, and to save the parish money*, 3 vols (Caistor, 1821)

Henthorn, F., *The History of Brigg Grammar School* (Brigg, 1969)

Holderness, B.A., 'The English land market in the eighteenth century: the case of Lincolnshire', *Economic History Review*, 2nd Series, 27 (1974), 557–76

Keelby, Parish and People 1765–1831 (Keelby, 1986)

Lincoln, Rutland and Stamford Mercury

Lincolnshire Archives Committee, *Archivists' Reports 1–28* (1948–83)

Morris, Jeff, *The Story of the Mablethorpe and North Lincolnshire Lifeboats* (R.N.L.I., 1989)

Nicholson, S.W., and Boyden, Betty, *The Middling Sort: The Story of a Lincolnshire family 1730–1900* (printed Nottingham, 1991)

Obelkevich, James, *Religion and Rural Society: South Lindsey 1825–1875* (Oxford, 1976)

Olney, R(ichard) J., *Lincolnshire Politics 1832–1885* (Oxford, 1973)

——, *Rural Society and County Government in Nineteenth-century Lincolnshire* (Lincoln, 1979)

——, 'The Enclosure of Holton-le-Moor', in Tyszka, Dinah, Miller, Keith, and Bryant, Geoffrey (eds), *Land, People and Landscapes: Essays on the History of the Lincolnshire Region written in honour of Rex C. Russell* (Lincoln, 1991), 121–4

——, 'The Youngs at Kingerby: The making, un-making and re-making of a Lincolnshire estate', in Sturman, Christopher (ed.), *Lincolnshire People and Places: Essays in Memory of Terence R. Leach (1937–1994)* (Lincoln, 1996), 117–20

——, 'Squire and community: T.G. Dixon at Holton-le-Moor 1906–1937', *Rural History* 18 (2007), 201–16

Padley, Christopher, 'Caistor Canal', *Lincolnshire History and Archaeology* 44 (2009), 5–22

Perkins, J.A., *Sheep Farming in Eighteenth and Nineteenth Century Lincolnshire* (Sleaford, 1977)

Rawding, Charles, *The Lincolnshire Wolds in the Nineteenth Century* (Lincoln, 2001)

——, 'Society and place in nineteenth-century North Lincolnshire', *Rural History* 3 (1992), 59–85

Richardson, Mary E., *The Life of a Great Sportsman (John Maunsell Richardson)* (London, 1919)

Russell, Rex C., *The History of the Enclosures of Nettleton, Caistor and Caistor Moors 1791–1814* (Nettleton, 1960)

——, *A History of Schools and Education in Lindsey, Lincolnshire 1800–1902, Part 2, The 'miserable compromise' of the Sunday School* (Lincoln, 1965)

——, 'Caistor Hospital' (unpublished typescript, 1973)
——, *Aspects of the History of Caistor 1790–1860, with special attention to the 1850s* (Nettleton, 1992)
—— and Holmes, Elizabeth, *Two Hundred Years of Claxby Parish History* (Claxby, 2002)
—— and Russell, Eleanor, *Making New Landscapes in Lincolnshire: The Enclosure of Thirty-four Parishes in Mid Lindsey* (Lincoln, 1983)
Saunders, David, *Caistor's Vicar and Church 1886–1916: the Rev Canon W.F.W. Westbrooke* (Heighington, 2003)
——, *More Portraits of Caistor, Lincolnshire: A Church, Doctors and Solicitors, Mills and an Inn* (Heighington, 2007)
——, *The Story of Caistor Grammar School Lincolnshire from 1631 to 1945* (printed, 2007)
Stovin, Jean (ed.), *Journals of a Methodist Farmer 1871–1875* (London, 1982)
Tyszka, Dinah, 'Clergy, church and people 1790–1860', in Russell, Rex C., *Aspects of the History of Caistor 1790–1860* (Nettleton, 1992)
White, William, *History, Gazetteer and Directory of Lincolnshire* (Sheffield, 1842, 1856, 1872, 1892)

General

Bell, Colin R., *Middle Class Families* (London, 1968)
Burke's Landed Gentry, 15th edn (1937)
Cannadine, David, *Class in Britain* (New Haven and London, 1998)
Davidoff, Leonore, and Hall, Catherine, *Family Fortunes: Men and Women of the English Middle Class 1780–1850* (London, 1987)
Earle, Peter, *The Making of the English Middle Class: Business, Society and Family Life in London 1660–1730* (London, 1989)
Eden, Sir Frederick Morton, Bt, *The State of the Poor* (London, 1797)
Everitt, Alan, *Landscape and Community in England* (London and Ronceverte, 1985)
Gerard, Jessica, *Country House Life: Family and Servants 1815–1914* (Oxford, 1994)
Hunt, Margaret R., *The Middling Sort: Commerce, Gender and the Family in England 1680–1780* (Berkeley and London, 1996)
Kent, J.R., 'The rural "Middling Sort" in early modern England c.1640–1740', *Rural History* 10 (1999), 19–54
Lewis, G.J., *Rural Communities* (London and Newton Abbot, 1979)
Lord, Evelyn, 'Communities of common interest: the social landscape of south-east Surrey 1750–1850', in Phythian-Adams, Charles (ed.), *Societies, Cultures and Kinship 1580–1850: Cultural Provinces and English Local History* (Leicester, 1993), 131–212
Martins, Susanna Wade, and Williamson, Tom (eds), *The Farming Journal of Randall Burroughes (1794–1799)* (Norfolk Record Society LVIII, 1995)
Morris, R.J., *Men, Women and Property in England 1780–1870: A Social and Economic History of Family Strategies amongst the Leeds Middle Classes* (Cambridge, 2005)
Newby, Howard, Bell, Colin, Rose, David, and Saunders, Peter, *Property, Paternalism and Power: Class and Control in Rural England* (London, 1978)
Orde, Anne (ed.), *Matthew and George Culley: Farming Letters* (Surtees Society 210, 2006)
Owners of Land, Return of (1873), *Parliamentary Papers,* 1874 lxxii

Phythian-Adams, Charles, *Re-thinking English Local History* (Leicester, 1987)
Read, Donald, *The English Provinces c.1760–1960: A Study in Influence* (London, 1964)
Rollinson, David, *The Local Origins of Modern Society: Gloucestershire 1500–1800* (London and New York, 1992)
Rowe, D.J., 'The Culleys, Northumberland farmers 1767–1813', *Agricultural History Review* 19 (1971), 156–74
Seed, John, 'From "Middling Sort" to "Middle Class" in late eighteenth and early nineteenth-century England', in Bush, M.L. (ed.), *Social Orders and Social Classes in Europe since 1500: Studies in Social Stratification* (London and New York, 1992)
Smail, John, *The Origins of English Middle-class Culture: Halifax, Yorkshire 1660–1780* (London, 1994)
Snell, K.D.M., *Parish and Belonging: Community, Identity and Welfare in England and Wales 1700–1950* (Cambridge, 2006)
Webb, Sidney and Beatrice, *English Poor Law History: Part One, The Old Poor Law* (London, 1927)
Woodruff, Douglas, *The Tichborne Claimant: A Victorian Mystery* (London, 1957)

INDEX

Places are in the pre-1974 County of Lincoln unless otherwise indicated.

accounts and accountants 45n, 50, 50n, 62, 106, 106n, 186–7; *see also* Plates 5, 12, 13
address, forms of 26, 41, 58
Adlingfleet, Yorkshire 18, 19
   tithes 18, 19
agents, land, *see* bailiffs and agents
agriculture, *see* farms and farming
Alford, School of Industry 71
Althorpe 109
Allenby family, of Cadwell, farmers and landowners 161, 173n
Ancaster, Duke of, *see* Bertie
Ancholme, river 6, 7, 47, 125
   Drainage and Navigation 25, 35, 104
Anderson-Pelham family, *see* Pelham
Andrews, Samuel Wright, rector of Claxby with Normanby 139
archive, Dixon family xi, 62, 152, 153, 185–90
aristocracy, titled 1, 2, 64, 65, 76; *see also* gentry, landed
Atkinson family, of Hatfield, Yorkshire, and north Lincolnshire
   Joseph Robert, of Binbrook Hill, farmer 103, 110
   Richard, of Hatfield, solicitor 109, 129n
   Richard, rector of Claxby with Normanby 110
   Susanna, *see* Dixon, Susanna
attorneys-at-law, *see* solicitors
Aylesby 31, 55, 74
Ayscough family, of South Kelsey 12

bailiffs and agents 20, 104, 113, 130, 136, 138n, 146–7; *see also* Burcham, John; Dudding, John; Marris, George; Mountain, Joseph; Seagrave, Henry; Slater, Charles; Todd, W.H.; Vessey family
Barkworth family, of Holton-le-Moor and Nettleton, farmers and small landowners 24, 45; *see also* Map 2
Barnetby-le-Wold 156n, 162
Barnoldby-le-Beck 22, 164n
Barton-on-Humber 8, 19, 31
Bayons Manor, Tealby 138, 182
Beauclerk, Lord William, of Redbourne Hall 56, 65
Beelsby 103, 161
Beltoft 129, 130, 140
Bennard family, of Owmby near Searby, farmers and landowners 161, 162n
Bertie, Brownlow, 5th Duke of Ancaster, Lord Lieutenant of Lincolnshire 39, 64
Bestoe family, of Holton-le-Moor 12–14, 25, 102
Bett family, of Holton-le-Moor, farmers and small landowners 32, 177
   Robert 61, 102
Binbrook 103, 110
Biscathorpe 162
Bishop Bridge Turnpike 23, 35
Borman family, of Irby and Swallow, farmers
   Elizabeth, of Swallow 131
   Thomas Johnson, of Swallow 135, 135n, 145, 148
   William, of Irby 110n
Boucherett family, of Willingham House 138
   Ayscough 100, 179n
Bower family, of Caistor

Anthony, headmaster of the Grammar
    School 131–2
  George, engineer 149n
Bowstead, Rowland, schoolmaster 50n,
  75, 76, 76n, 118
Bradley Haverstoe wapentake 72
Brandy Wharf, Waddingham 103, 126n
Brattleby 163, 166
Bridlington, Yorkshire 139
Brigg 7, 8, 31, 38
  Grammar School 32
  market area 46, 74, 178
Briggs family, of Oxcombe, farmers and
    landowners 161, 166
Brocklesby 38, 118
  estate 8, 9n, 28, 29, 36, 43, 45n, 72, 73,
    132, 148, 156, 157–8, 160, 162,
    163
  Hunt 9n
  see also Pelham family
Bromhead family, Baronets, of Thurlby
    near Lincoln 139
Brooks family, of Croxby, etc, farmers and
    landowners 156n, 157
  Thomas, of Hundon 131
Broughton family, of Market Rasen and
    Holton-le-Moor 17n, 25, 177
  Ann, *see* Dixon, Ann (born Ann
    Waltham)
  Benjamin 14, 17, 32, 101–2
  Sailbanks (d. 1731) 17
Brown, Edward, of Holton-le-Moor,
    farmer 117
Brownlow, Earl, *see* Cust
buildings, domestic, *see* cottages;
    farmhouses; houses, middle-class
    and gentry
buildings, estate and farm 30n, 44, 46,
    116, 126, 128, 130, 147; *see also*
    farmhouses
Burcham, John, of Coningsby, land
    agent 49
Burroughs family, of Norfolk, farmers and
    landowners 168
Byron family, of North Killingholme,
    farmers and landowners 132,
    143–4

Cadwell 173n
Caistor 6, 7, 8, 8n, 9, 40, 125, 126, 137,
    178, 184
  Association for the Prosecution of
    Felons 41
  bank 109n
  Canal 46, 47n, 54, 69, 125
  Dixon family connections 20, 30–1, 41,
    71–7, 103, 108–11, 119, 137, 145,
    178, 179
  doctors, *see* Jameson, George; Porter,
    George Markham; Turner, Samuel
  fair 7, 20, 33, 50
  Friendly Society 76, 78, 105
  gas company 132
  George Inn 72, 73, 76, 77
  Grammar School 50, 99, 179n
  Holly House 43, 144, 145
  malting business 103, 109
  manorial lordship 73
  market and social area 8, 20, 30, 34, 46,
    72, 137, 178–9
  Matron Society 77, 78
  Moor 56, 73
  National school 137
  Poor Law union 78, 180; *see also
    below,* Society and House of
    Industry
  prebend 56
  quarter sessions 40
  Savings Bank 77, 78
  Skipworth property 55, 69, 153
  Society and House of Industry 71–8,
    180
  solicitors 51–2; *see also* Dixon,
    Marmaduke; Marris and Smith
  Sunday school 76, 105
  tannery 103, 103n, 109, 130
  Tower House 51, 51n, 111
  White House 108, 111, 119, 152
Carlton, Great 21
Carlton, North 162, 162n
carriages 37, 43, 58, 118; *see also*
    coachmen and grooms
Cary family of Owersby and Grasby,
    farmers and landowners 29, 31, 38
Cave, Christopher, schoolmaster 32, 50

Charles, Anthony, of Nettleton,
    accountant 127, 127n
Chatterton family, of Stenigot, farmers and
    landowners 161
Cholmeley, Sir Montague, Bt, of Norton
    Place 75, 76, 78, 78n
class, social 1–2, 6, 64–8, 170, 171–6; *see
    also* address, forms of; aristocracy,
    titled; deference; gentry, landed;
    labourers, agricultural; middle class
    or middling sort
Claxby-by-Normanby 6, 17, 137, 148, 177
    living and glebe 36, 44, 45
Cleatham 103, 163, 167
Cleethorpes 103, 118, 139
clergy 3, 4n, 28, 35–6, 37–9, 66, 69, 119,
    139, 145–6, 150, 173
coachmen and grooms 118, 135, 136, 150,
    174–5
Coates family, of Beelsby and Hatcliffe,
    farmers and landowners 103, 161,
    161n
Collinson family, of Beltoft
    John, solicitor 129, 129n
    Lucy, *see* Dixon, Lucy
Collyer Bristow, of London, solicitors, *see*
    Eyre and Coverdale
Colton, Richard, of Moortown,
    builder 116
Commission of the Peace 39–40, 119–20,
    140, 141, 161, 172–3
communities, local xii, 5–6, 66, 68–71,
    176–81, 182; *see also* markets and
    market areas; *pays*
Cook family, of Holton-le-Moor, farmers
    and innkeepers
    Robert 124
    Tom 148, 151
Cooper family, of West Rasen, clergymen
    William 120
    William Waldo 146
Cotes family, of Normanby-by-Spital
    and Walesby, graziers and
    landowners 21
    John 21, 22, 23
cottages 11, 66, 67, 116, 117n, 128, 130,
    148–9

Coverdale, John, of London, solicitor 129,
    132, 133, 134, 135; *see also* Eyre
    and Coverdale
Croxby 37n, 157
Crust, Thomas, of Beverley, Yorkshire,
    solicitor 146; *see also* Crust, Todd
    and Mills
Crust, Todd and Mills, of Beverley,
    Yorkshire, solicitors 146, 186; *see
    also* Crust, Thomas; Mills, James;
    Todd, William
Culley family, of Northumberland, farmers
    and landowners 167–8
Cust, John, 1st Earl Brownlow, Lord
    Lieutenant of Lincolnshire 119–20

Dauber family, of Brigg and Ruckland
    John, corn merchant 112
    see *also* Dixon, Elizabeth
Dawson, Richard, of Withcall, farmer 156
deference 5, 69, 117, 117n; *see also* class,
    social
depression, agricultural 64n, 71, 102, 145,
    147–8, 152, 166, 177
Dixon family, of Holton-le-Moor
    Amelia Margaretta (born Parkinson, m.
        William Dixon) 33, 38, 47–8, 49,
        58, 58n, 77, 104, 105, 106, 123,
        138
    Amelia Margaretta (m. George
        Skipworth), *see* Skipworth
    Amelia Margaretta, Mrs Jameson Dixon
        (m. Dr George Jameson) xii, 129,
        130, 138, 188; *see also* Plates
        15(a), 16
        birth 115
        death and will 145, 150
        estate policy 147–9
        family life 130, 141
        inheritance 143, 144–5
        marriage 144
        religious views 150–1
        social position 150–1, 176
    Ann (b. Waltham, m. (1) Thomas Dixon,
        (2) Sailbanks Broughton) 15, 17
    Ann (m. Richard Roadley), *see* Roadley
    Ann (d. 1893) 129, 138, 141, 150

birth and education 115, 129
charitable work 151
death and will 144–5, 152
estate policy 147, 148
family life 130, 141
inheritance 134, 143, 144
social position 150, 176
Charlotte Roadley 115, 129; *see also* Plates 7(a), 16
Elizabeth (b. Dauber, m. James Green Dixon) 108
Frances and Martha, of Deloraine Court, Lincoln, daughters of the Revd Richard Dixon 139
George Sperling (previously Gibbons) xii, xiii, 188–9
James Green, of Caistor, landowner, farmer and merchant 47, 50, 52, 57, 58, 78, 104–6, 108–9, 111, 119, 131, 137
James Green junior, of Caistor 152, 188
Jane, *see* Kelk
John William, of Thornton-le-Moor and Caistor 131, 132, 134, 152, 153
Joseph, of Buslingthorpe 16, 29
Lucy (b. Collinson, m. Richard Roadley Dixon) 129, 143
Marmaduke, of Fulbeck 16
Marmaduke, of Caistor, solicitor and landowner 47, 50, 51–2, 75, 77, 102–11
Marmaduke, of Eyrewell, New Zealand 134
Martha (b. Walkden, m. Thomas Dixon) 28, 48, 49, 52, 54, 57–8, 76
Mary, *see* Smith
Mary Ann (b. Roadley, m. Thomas John Dixon) 106–7, 117, 129, 133, 133n, 135–6, 141, 143–4; *see also* Plate 11
Mary Ann (d. 1856) 115, 129, 133
Rachel (b. Drewry, m. William Dixon) 18, 21
Rachel, *see* Etherington
Richard, rector of Claxby 32, 36, 36n, 54

Richard Roadley 115, 120–1, 128–30, 133–4, 140–1, 143, 151
Robert (d. 1593), of Owersby 14
Robert (d. 1627), of Owersby 15
Susanna (b. Atkinson, m. Marmaduke Dixon of Caistor) 109–11
Thomas (d. 1704) 15, 16
Thomas, of Riby, farmer and landowner 16, 17, 22, 27–41 *passim,* 169–70
birth and early life 20–1
death and will 41, 47–9
family affairs 32–3, 35–6, 38–9, 44
farming career 22, 27–8, 30–1, 33, 35
financial and estate affairs 33, 35
local affairs and public office 39–40, 187
marriage and birth of children 28, 32
social position 37–41, 174
Thomas, rector of Laceby 32, 35–6, 54
Thomas George (previously Gibbons) xii, 78, 134, 145, 152, 153, 184, 185, 188
Thomas John, farmer and landowner
birth and education 43, 50
business enterprises 103
death and wills 120–1, 133–5, 141, 143
estate policy 127–8, 130–1
family affairs 104–6, 107–8, 110–11, 128–30, 138
farming career 50–1, 56, 57, 100, 103, 107, 111–15, 123, 125–7, 160n, 165
financial affairs 111, 112, 123–4, 131–2, 143, 175
local affairs and public office 77, 103–4, 119–20, 140
marriage and birth of children 107, 115
purchases of land 101–2, 109, 123, 132
record-keeping 186–7
social position 115–18, 127–8, 135–6, 138–9
see also Plate 10
Thomas John (d. 1855) 115, 121, 128, 129, 133

Thomas Parkinson 152
William, of Middle Rasen 15
William (d. 1782), grazier and
    landowner 11–26 *passim*, 29, 31
  birth and early life 15, 17
  death and will 26, 32
  farming and grazing career 17–26
  marriage and birth of son 17–18, 20
  public office 20, 22
  purchases of land 11, 12, 19, 21,
    23–5
  wealth and social position 20, 25–6
William (d. 1824), farmer, landowner
    and philanthropist 43–8 *passim*
  birth and early life 26, 32, 43, 61
  death and will 104–6
  family affairs 33, 35, 38, 47–9,
    48–53
  farming career 26, 32, 44–7, 51, 52–4
  financial affairs 44, 45, 47, 51, 52–4,
    56, 105
  local affairs and public office 71–7
  marriage and birth of children 33, 38,
    43, 44, 47
  purchases of land 44–5, 51, 181
  record-keeping 185
  religious and social opinions 58–9,
    61–71
  social position 43, 44, 58, 64, 69,
    174–5
doctors of medicine 3, 4, 139, 173; *see
    also* Jameson, George; Porter,
    George Markham; Turner, Samuel
Donna Nook lifeboat 151
Drewry family, of Adlingfleet,
    Yorkshire 18
  Rachel, *see* Dixon, Rachel
Dudding, John, farm bailiff 113
Dudding, John Walter, of Saxby and
    Howell 108, 182n

Eardley (sometime Eardley-Smith) family
  estate in Nettleton 114, 132, 156
  Sir Culling 78n
  Sampson, Baron Eardley 76
education 17, 20, 32, 47, 55, 67, 75, 76–7,
    128–9, 172, 175

Egginton family, of Hull and North
    Ferriby, Yorkshire, merchants
  Charlotte (born Roadley) 106–8, 145
  Samuel Hall 107
enclosure 13, 22, 22n, 30n, 45–6, 51,
    56–7, 67
estate management 127, 130–1, 132, 136,
    145–7; *see also* bailiffs and agents
Etherington family, of Gainsborough,
    merchants
  Rachel (born Dixon, m. William
    Etherington) 32, 38, 48
  William 38, 47, 59
executorships, *see* settlements and wills
Eyeworth, Bedfordshire 36
Eyre and Coverdale (later Collyer
    Bristow), of London,
    solicitors 110, 153, 186; *see also*
    Coverdale, John

fairs 7, 31, 33, 50, 100
farms and farming 1, 4, 19, 27–8,
    30–1; *see also* accounts and
    accountants; buildings, estate and
    farm; depression, agricultural;
    Dixon family *passim*; farmers and
    graziers; farmhouses; farm sizes;
    labourers, agricultural; landlord-
    tenant relations; servants, farm
farmers and graziers
  and landownership 2, 155–70
  as tenants 2, 64, 130–1, 132, 147,
    157–8, 160, 169–70
  retirement 26, 41, 56, 172
  see also landlord–tenant relations;
    middle class or middling sort
farmhouses 11, 28, 30, 30n, 43, 128n, 130,
    147, 177
farm sizes 2, 19, 20, 22, 27, 30, 46, 51,
    53, 56, 57, 64, 103, 111–12, 123,
    158–60, 169, 172
Favill, Robert, farm foreman 127
Ferraby, John, of Owmby (by Searby)
    and later of Wootton, farmer and
    landowner 105, 133, 143
Field family, of Laceby 35
  David, rector of Thornton-le-Moor 103

Firsby, West 22, 27–8
Folkingham 71
Frankish, William, of Great Limber, farmer 157
French Wars, agricultural prosperity during 1, 63, 66, 163, 171, 183–4

Gainsborough 7, 38, 40
gentry, landed 6, 29, 57, 75–6, 78, 119, 167, 168, 175–6, 181–2; *see also* address, forms of
Gibbons family, of Worcestershire and later Holton-le-Moor xii
   Benjamin, of Waresley, Worcestershire 134, 134n
   Charlotte Jane (born Skipworth, m. Benjamin Gibbons) 134, 144, 145
   George Sperling, *see* Dixon
   Thomas George, *see* Dixon
Goodson, William, of Market Rasen and Holton-le-Moor, innkeeper and farmer 130, 130n
Goulton family, of Croxby, farmers 37, 37n, 46, 157
Goxhill 64
Grant family, of Hareby, farmers and landowners 161
Grant, John, of Oxcombe, farmer and landowner 166
Grimsby, Great 6, 8, 31

Hargrave family, of Holton-le-Moor and later Caenby, farmers and landowners 148
Harrogate, Yorkshire 106, 107, 108, 118, 139
Healing 33, 38, 74, 164, 164n, 165
Hebb, H.K., of Lincoln, solicitor 146; *see also* Hebb and Sills
Hebb and Sills, of Lincoln, solicitors 146, 186
Hewitt family, of Holton-le-Moor, small farmers 117, 177
   Thomas 136
   William 69
Hibaldstow 38

Hill family, of Winceby, farmers and landowners 161
Holgate, William, of Keelby, farmer 73
Holiwell, Revd George, of Irby 37, 75
Holton-le-Moor
   almshouses 149
   Barkworth's farm 14, 24, 44, 126; *see also* Hall Farm
   Bestoe house 11–13
   Bett's farm 102
   brickyard 114, 124, 127
   Broughton's house and land 11n, 14
   church 11, 13, 61, 117, 128, 136, 150; *see also* Plates 4(a), 4(b)
   communications 6–7, 8; *see also* Map 1
   cottages 11, 11n, 116, 117n, 128, 148–9
   Daisy Hill (or Grange) Farm 13, 13n, 114, 117, 128, 143, 149; *see also* Map 2
   enclosure 13, 13n, 56–7, 114
   Ewefield Farm 13, 19, 20, 21, 32, 33, 44, 46, 53, 112, 114, 130, 131, 147, 148; *see also* Map 2
   farms and farmhouses 11, 13, 130, 147; *see also* Map 2
   farm buildings 46, 53, 116, 130, 147
   Hall 33, 43–4, 58, 115–16, 185–6; *see also* Plates 1, 2
   Hall, Home or Manor Farm 12, 19, 24, 33, 46, 48, 54, 56, 103, 127, 130, 146, 147
   Hope Tavern 124, 127, 137, 151
   Jacklin's farm 25, 33, 44, 66; *see also* Map 2
   manor house, former 11, 12, 14, 22, 127, 149; *see also* Plate 3(a)
   manorial lordship 12, 19
   Maze 13, 24; *see also* Map 2
   Moor 6, 11, 13, 56, 114–15; *see also* warren; Map 2
   Mount Pleasant Farm 45, 46, 52, 53, 54, 66, 104, 105, 130, 131, 146, 148; *see also* Map 2
   Noble's 25
   nonconformity 136–7
   parish and parochial offices 7, 20
   park 116

plantations 114, 125, 126, 127
population 6, 11, 150
railway and railway station 124–5, 128, 148, 150, 177
school 61, 76, 105, 117, 128, 136, 147, 150
Stope Hill 46, 51, 125, 126, 130, 133, 147
Taylor's Farm 33
tithes 56, 112, 114–15, 128
village 11, 136–7, 177
warren 6, 11, 12, 13, 14, 14n, 24, 32, 33, 45, 46, 48, 54, 56–7, 114, 114n
water supply 149
Wiles's farm 44–5
Yewfield, *see* Ewefield
Horncastle 31, 100
houses, middle-class and gentry 165, 174, 174n; *see also* Bayons Manor; Caistor, Holly House; Holton-le-Moor, Hall; Kingerby, Hall; Moortown House; Owmby Mount; Searby, Manor
Hudson, Benjamin, portrait painter 116
Hudson family, of Kirmington, farmers and landowners 156n, 161, 162, 167, 167n
Hull, Yorkshire 31, 103, 108, 113n, 114n, 125
Humberstone 39, 55n
Humble, Edward, of Renishaw, Derbyshire 112n

Iles family
   John, farmer, of Binbrook Hill 135, 135n
   John Cyril 146
   Ven John Hodgson 145, 150
inheritance 3, 18, 162–3, 165–6, 168; *see also* settlements and wills

Jackson family 108
   William, rector of Nettleton 39, 39n, 50
Jameson, George, of Caistor, doctor 141, 144, 145

Keelby 28, 31, 73

farm in 34, 35, 44, 45, 48, 48n
proposed workhouse 72
Sunday school 76
Kelk family, of Brigg, merchants and farmers 34, 38, 139
   Jane (born Dixon, m. John Kelk junior) 32, 38, 48, 51
Kelsey, North 55, 76, 126, 133, 134
Kelsey, South 6, 12
   Skipworth estate 55, 152–3, 163, 166–7; *see also* Moortown
Killingholme, North 132, 164n
Kingerby 165, 166, 167n
   Hall 131, 165
Kirkby-cum-Osgodby 15
Kirkby, John, of Caistor, genealogist 188
Kirmington 156n, 162
Kirton-in-Lindsey 40, 180

labourers, agricultural 46, 65, 66–8, 70, 126, 136, 177, 187; *see also* servants, farm
Laceby 31, 74, 164n
   Free School 32
   living 35
landlord–tenant relations 4, 65, 157
land market, rural, *see* land purchase and ownership
landowners, *see* aristocracy, titled; gentry, landed
land purchase and ownership 2, 4, 21, 21n, 19–25 *passim*, 44–5, 55, 57, 65–6, 101–2, 108–12, 123, 128, 132–3, 143, 155–70
Lawrence family, of Caistor
   Thomas 73, 73n
   Mrs Lawrence, widow of Thomas 69, 77
Leslie-Melville family, of Lincoln, bankers 139
Limber, Great 28, 40
   Boundary Farm 43, 43n
Lincoln 8, 139, 180
   assizes 41, 140
   charitable institutions 140, 151
   Dean and Chapter 23, 49n, 128, 143n
   fair 33, 34, 50
   Poor Law union 72

London 39, 51, 59, 100, 115, 118, 124, 125, 139, 175, 180
Lord, William, of Holton-le-Moor, accountant and sub-agent, 127, 151, 188
Louth 7
  market area 8, 9n, 24

Mablethorpe 103
Mackenzie (previously Humberston) family, of Humberstone and Somerby near Gainsborough 39, 39n
Maddison family, of Holton-le-Moor 117, 150
  David, coachman 102n, 136, 141
  George, small farmer 127, 136
  Jane, wife of David 136
Madley, Herefordshire 101, 112
magistrates, *see* Commission of the Peace
Manchester, Sheffield and Lincolnshire Railway, *see* railways
markets and market areas 7, 8–9, 15, 30, 68, 70, 74, 100, 178; *see also* Barton- on-Humber; Brigg; Caistor; Grimsby, Great; Louth; Rasen, Market
Marris family, of Great Limber, farmers 110n
  George, of Caistor, solicitor 106, 110, 110n, 118, 135n, 138n
  Thomas, of Ulceby Grange, farmer 135n
Marris family, of Thoresway, farmers
  Eliza, *see* Seagrave, Eliza
  George, of Holton-le-Moor, farmer and agent 147
Marris and Smith, of Caistor, solicitors 146, 186; *see also* Marris, George, of Caistor; Smith, Charles
Marris, Thomas, of Barton-on-Humber 73
Maw family, of Bigby and Brigg, later of Cleatham 110, 161, 163, 167
  George, of Bigby, farmer 163
  Matthew, of Brigg, corn merchant 163

Matthew, of Cleatham, farmer and landowner 103, 173n
merchants 3, 4, 12, 38, 46, 68, 100, 104, 108–9, 137, 173, 174
Messingham 38, 106, 107
Methodism, *see* nonconformity
middle class or middling sort xi, 1–5, 37–8, 66, 73, 137, 174–5, 181; *see also* clergy; doctors; farmers; merchants; solicitors
Mills, James, of Beverley, Yorkshire, solicitor 146–7; *see also* Crust, Todd and Mills
Moortown (South Kelsey) 47, 100, 125
  Moortown House 55, 118, 135
Monson family, Barons Monson 14, 162
  Owersby estate 14, 19, 29
Mountain, Joseph, of Holton-le-Moor, bailiff 113, 127, 131

Napoleonic Wars, *see* French Wars
neighbourhoods, *see* communities, local
Nelson family, of Great Limber and Wyham-cum-Cadeby, farmers and landowners 46, 57, 132, 144, 161, 167n
Nelthorpe family, Baronets, of Scawby 36n
  North Kelsey property 133, 156
Nettleton 6, 39, 74, 126, 137, 177
  Bleak House or Prospect Farm 128
  Dixon property and farms 45–6, 50–1, 100, 101, 102, 128, 131, 132, 134
  Eardley estate 114, 132
  enclosure 45–6, 51
  ironstone mine 150
  living and glebe 50–1
  New Farm 132
  Oxgangs Farm 132
  South Moor Farm 132
  Wold Farm 132
  see also Holton-le-Moor, Stope Hill
Newton-by-Toft 15
Newton, Wold 160, 162, 167
Nicholson family, of Laceby and Keelby, later of Wootton, farmers and landowners 31, 156n

Brady 99, 101n, 119
  Elizabeth 101
Noble family, of Holton-le-Moor, small farmers 40, 117, 177; *see also* Map 2
nonconformity 63, 66n, 131, 136–7
Normanby-by-Spital 16, 21
  Dixon property 23, 34, 35, 48
Normanby-le-Wold 6, 7, 19, 177

offices, local 20, 39, 69, 71, 103, 119, 140, 172, 175, 181; *see also* Commission of the Peace
Otter family, of Ranby 161, 163n, 167, 173n
  Francis, of Stainton-le-Vale, farmer and landowner 163, 163n
Owersby 13n, 14, 18, 19, 22–3, 29, 31, 177
  Angerstein estate cottages 148–9
  Dixon farm 19, 22, 26
  Foresters 137, 141
Owmby (near Searby) 162, 162n
  Dixon property 133, 134
  Owmby Mount 143, 144
  see also Searby-cum-Owmby
Owmby-by-Spital 16
  Dixon family of 16–17
Oxcombe 162, 163n, 166, 166n

Parkinson family, of East Ravendale 38, 110, 119, 161, 163–4, 165, 166, 167, 179; *see also* Table 2
  John, rector of Brocklesby 38
  Ursula, *see* Wright family of Brattleby
Parkinson family of Healing, clergymen, farmers and landowners 38, 119; *see also* Table 2
  Amelia Margaretta, *see* Dixon, Amelia Margaretta (born Parkinson)
  Frances (born Green, m. John Parkinson) 101, 108n
  John, rector of Healing 33, 38
  Robert, of Barnoldby-le-Beck, farmer and landowner 35, 38, 47, 73, 101, 105, 108
pays 8–9

Pelham, later Anderson-Pelham, family, Barons, and later Earls of, Yarborough
  Charles Anderson, 1st Baron 36, 39, 65, 65n, 74, 76, 77, 179
  Charles Anderson Worsley, 2nd Earl 120, 179
  Victoria, widow of the 3rd Earl 138, 176n
  local influence 8, 28, 65, 72, 74, 77, 119, 179
politics, party 119–20, 179; *see also* Pelham family
Poor Laws 70–5, 78; *see also* Caistor, Society and House of Industry
Porter, George Markham, of Caistor, doctor 118
portions, marriage, *see* settlements and wills

Quirk, John Francis, vicar of Great Coates 146, 150

rabbits, *see* warrens
railways 7, 8, 124–5, 132, 139, 178, 181
Ranby 163, 167
Rasen, Market 6, 7, 8, 178
  Grammar School 17
  market area 8, 15, 20, 28, 30, 74, 178
  yeomanry and volunteers 100, 119, 140, 141
Rasen, Middle 15, 16
Rasen, West 15, 165
Ravendale, East 38, 74, 164
Raynes family, of Everton, Nottinghamshire, 53n, 69n, 104, 163n
religious attitudes, 36, 61–3; *see also* clergy; nonconformity
Riby 29, 39, 74
  Dixon farm 30, 49
  Tomline estate 16, 29, 31, 49, 156; *see also* Tomline family
Richardson family, of Great Limber, farmers and landowners 40, 180
  Thomas Martinson, of Hibaldstow 135, 135n

212　INDEX

William 73, 157n, 176n
Riverhead, *see* Moortown
Roadley family, of Messingham
    and Searby, farmers and
    landowners 38–9, 106, 161; *see
    also* Table 3
  Ann (born Dixon, m. Richard
    Roadley) 32, 38–9, 48, 49, 54, 59,
    77, 106, 107, 137, 138
  Charlotte, *see* Egginton, Charlotte
  Mary Ann, *see* Dixon, Mary Ann (m.
    Thomas John Dixon)
  Richard, of Riby and later Searby 38,
    48, 49–50, 50–1, 73, 99–100, 103,
    105, 106, 112
  Richard Dixon 106
Rothwell 22, 24, 55, 108
  James Green Dixon's farm 108, 152
Rothwell family, of Thorganby 14, 24

Saunderson family, of Nettleton, farmers
    and stock-jobbers 33, 34
Scarborough, Yorkshire 129, 139
Scunthorpe, Parkinson property at 164,
    165
Seagrave family, of Lissington
  Elizabeth (born Marris, m. Henry
    Seagrave) 131, 147
  Henry, of Holton-le-Moor and later
    Kingerby, bailiff and agent 127,
    130, 131, 146, 149, 163
  William, of Lissington, agent to George
    Skipworth 138n
Searby-cum-Owmby 22, 74, 125, 137,
    164, 167n
  church 118
  Manor 49, 137, 148, 174n
  Roadley, later Dixon, estate and
    farm 49, 49n, 57n, 106–7, 112–13,
    118, 120–1, 126, 128, 131, 133,
    140, 143–5, 148, 161, 167
servants, domestic 26, 31, 37, 40, 43, 58,
    115, 135–6, 150, 174, 175
servants, farm 31, 37, 40, 41, 46, 66, 70,
    113
settlements and wills 17n, 20, 21, 26,
    29, 33, 33n, 35, 44, 47–9, 54,
    56, 104–6, 107, 109, 120–1,
    129, 133–5, 143–5, 157n,
    166–7, 176
shares, railway, *see* railways
shrievalty, *see* offices, local
Skidbrook, Dixon land in 21, 151
Skipworth family, of South Kelsey 55,
    161, 166–7, 180; *see also* Table 4
  Amelia Margaretta (born Dixon, m.
    George Skipworth) 47, 54, 138
  Ann Elizabeth, *see* Andrews
  Charlotte Jane, *see* Gibbons
  George, of Moortown House 54–5, 105,
    106n, 108, 109, 110, 111, 119–20,
    138, 138n, 152–3, 181
  George Borman 132, 138, 153, 173n
  Henry Green, of Rothwell, farmer 131
  Marmaduke Parkinson 134n
  Philip, of South Kelsey 55, 65, 69, 73,
    163, 163n, 182–3
  Philip, of Laceby 157n
  Septimus Patrick, of Owmby
    Mount 144
  Thomas, of Cabourne 101n
  Thomas, vicar of
    Belton-in-Axholme 109n
  Thomas, of Riby 29n, 31, 55
  William, of South Kelsey 106, 108,
    110n, 118, 129, 135, 145
Slater family, of North Carlton, farmers
    and landowners 161, 162, 162n
Slater, Charles, farm bailiff 113
Smith, Charles, of Caistor, solicitor 110,
    110n; *see also* Marris and Smith
Smith, Mary (born Dixon, m. R. Smith of
    Skipton, Yorkshire) 32, 38, 48
solicitors 3, 4, 51–2, 109–11, 138, 146,
    174, 186; *see also* Crust, Todd and
    Mills; Dixon, Marmaduke; Eyre
    and Coverdale; Hebb and Sills;
    Marris and Smith; Tennyson and
    Main
Somerby, near Gainsborough 39
Sowerby family, of Beelsby and Hatcliffe,
    farmers and landowners 103, 161,
    161n
  Thomas, of Withcall, farmer 161n, 165

Spilsby Society of Industry 71
Spridlington 16
squirearchy, *see* address, forms of; gentry, landed
Stenigot 163n
Storr, William, of Stope Hill, farmer 130, 136, 146
Stothard, Samuel, of South Kelsey, farmer 103
Stovin, Cornelius, of Hirst Priory, Crowle 24, 25, 44, 45, 45n
Stovin, Cornelius, of Binbrook, farmer 63n, 67n

Tennyson, later Tennyson D'Eyncourt, family, of Tealby 56n, 118, 138
    Charles Tennyson D'Eyncourt, of Bayons Manor, Tealby 56n, 118, 120, 182
    George, solicitor and landowner 51, 55, 55n, 73, 109n, 182
    Mary (born Turner, m. George Tennyson) 77
Tennyson and Main, of Market Rasen, solicitors 51, 182; *see also* Tennyson, George
Theddlethorpe, Dixon land in 21, 102, 112
Thoresby, North 110
Thorganby 14, 24
Thornton-le-Moor 6, 137, 177
    Beasthorpe Farm 35, 48, 51, 100, 126, 128, 130, 131, 133, 134, 145, 147
    church and living 53, 103, 119, 139
    Dixon estate in 23, 34–5; *see also* Beasthorpe Farm; Gravel Hill Farm
    Gravel Hill Farm 35, 48, 51, 52, 54, 104, 152
Thorpe, William, of Owersby, grazier 26
Todd, William, of Beverley, Yorkshire, solicitor 147
Todd, W.H., of Hull, Yorkshire, land agent 147
Tomline family, of Riby Grove
    George 156
    George Pretyman, later Pretyman-Tomline, Bishop of Lincoln 30n, 49, 75, 76

    Marmaduke 16, 29–30, 30n, 35, 41, 48n, 49
    William Edward 76, 77, 110
Torr family, of Riby, farmers 29n, 31, 46, 103, 106
    William 165
Tupling, Nathaniel, of Wrawby, brickmaker 116
Turner family, of Caistor 24, 46n
    John, of Caistor, attorney 45, 73
    Jonathan, of Riby, farmer 30
    Mary, *see* Tennyson
    Samuel, curate of Caistor 71–7 *passim*
    Samuel, of Caistor, doctor 73
    Samuel, rector of Nettleton 139
Tyberton, Herefordshire 101, 112

Ulceby (near Barton-on-Humber) 67
Usselby 15

Vessey family, of Holton Holegate and Welton-le-Wold, farmers, land agents and landowners 161, 161n

Walesby 21
Walkden family, of Great Limber
    Martha, *see* Dixon
    Richard, farmer 31, 34, 43
    Thomas, clergyman and farmer 28–9, 31
Warmer, Thomas, of Caistor, builder 43, 43n, 51n
Walshcroft wapentake 7, 72
warrens 14, 37n; *see also* Holton-le-Moor, warren
Wells-Cole family, of Fenton, etc, farmers and landowners 161, 162–3
Welton-le-Wold 161
Westbrooke, W.F.W., vicar of Caistor 78, 150
Weston family, of Somerby near Brigg 49n
Whitlam family, of Biscathorpe, farmers and landowners 161, 167
Whitworth, George, of Acre House, Claxby-by-Normanby, farmer 103
Wilkinson family, of Grasby

John (?son of William), of Barton-on-Humber, accountant 106, 106n, 127n
William, vicar of Grasby 73
Willoughby family, Barons Middleton 24–5
Willoughby family, Barons Willoughby de Eresby 156
wills, *see* inheritance; settlements and wills
Winship family, of Laceby and Riby 29n, 31, 35, 40
Withcall 156, 165
Wold Newton, *see* Newton, Wold
Wolds, Lincolnshire 6, 7, 8, 9, 9n, 159, 164, 179
Woolmer family, of Barton-on-Humber 35, 36n
Wootton 156n
Wragby 15
Wright family, of Brattleby 161, 164, 166, 167

Ursula (born Parkinson, m. William Wright) 164
William 163
Wright family, of Wold Newton, farmers and landowners 133, 160, 161, 162, 166, 167
William 162
Wyham-cum-Cadeby 132, 167n

Yarborough wapentake 7, 70, 72
coronership 52
Yarborough, Barons, and later Earls of, Yarborough, *see* Pelham
yeomanry and volunteers 41, 100, 119, 140, 141
Yorkshire 139, 178, 181; *see also* Beverley; Bridlington; Ferriby, North; Harrogate; Hull; Scarborough
Young family, of Kingerby 131, 165, 167n
Edward, of Normanby-le-Wold, farmer 42